Entropy Bounds and Isoperimetry

MEMOIRS
of the
American Mathematical Society

Number 829

Entropy Bounds and Isoperimetry

S. G. Bobkov
B. Zegarlinski

July 2005 • Volume 176 • Number 829 (first of 4 numbers) • ISSN 0065-9266

American Mathematical Society
Providence, Rhode Island

2000 *Mathematics Subject Classification.*
Primary 46–XX; Secondary 51–XX, 52–XX.

Library of Congress Cataloging-in-Publication Data

Bobkov, Serguei G. (Serguei Germanovich), 1961–
Entropy bounds and isoperimetry / S. G. Bobkov, B. Zegarlinski.
p. cm. — (Memoirs of the American Mathematical Society, ISSN 0065-9266 ; no. 829)
"Volume 176, number 829 (first of 4 numbers)."
Includes bibliographical references.
ISBN 0-8218-3858-X (alk. paper)
1. Isoperimetric inequalities. 2. Entropy (Information theory). 3. Inequalities (Mathematics). 4. Boundary value problems. I. Zegarlinski, B. II. Title. III. Series.

~~QA447.B567 2005~~
516–dc22 QA3 .A57 no. 829 2005045310

Memoirs of the American Mathematical Society

This journal is devoted entirely to research in pure and applied mathematics.

Subscription information. The 2005 subscription begins with volume 173 and consists of six mailings, each containing one or more numbers. Subscription prices for 2005 are $606 list, $485 institutional member. A late charge of 10% of the subscription price will be imposed on orders received from nonmembers after January 1 of the subscription year. Subscribers outside the United States and India must pay a postage surcharge of $31; subscribers in India must pay a postage surcharge of $43. Expedited delivery to destinations in North America $35; elsewhere $130. Each number may be ordered separately; *please specify number* when ordering an individual number. For prices and titles of recently released numbers, see the New Publications sections of the *Notices of the American Mathematical Society.*

Back number information. For back issues see the *AMS Catalog of Publications.*

Subscriptions and orders should be addressed to the American Mathematical Society, P. O. Box 845904, Boston, MA 02284-5904, USA. *All orders must be accompanied by payment.* Other correspondence should be addressed to 201 Charles Street, Providence, RI 02904-2294, USA.

Memoirs of the American Mathematical Society is published bimonthly (each volume consisting usually of more than one number) by the American Mathematical Society at 201 Charles Street, Providence, RI 02904-2294, USA. Periodicals postage paid at Providence, RI. Postmaster: Send address changes to Memoirs, American Mathematical Society, 201 Charles Street, Providence, RI 02904-2294, USA.

This publication is indexed in *Science Citation Index®*, *SciSearch®*, *Research Alert®*, *CompuMath Citation Index®*, *Current Contents®/Physical, Chemical & Earth Sciences.*
Printed in the United States of America.

♾ The paper used in this book is acid-free and falls within the guidelines established to ensure permanence and durability.
Visit the AMS home page at `http://www.ams.org/`

10 9 8 7 6 5 4 3 2 1 10 09 08 07 06 05

Contents

Abstract

We consider coercive and isoperimetric inequalities for probability measures with super-Gaussian tails. First part of the work is devoted to analysis of modified forms of logarithmic Sobolev inequalities, which we call LS_q, on abstract (metric) spaces with respect to L^q-norm of the modulus of gradient. We give an explicit characterization of probability distributions satisfying these inequalities on the real line. It is also shown that LS_q are satisfied for a large class of non-trivial infinite dimensional measures (such as Gibbs measures). In the second part, Sobolev-type inequalities are studied in the class of uniform distributions on convex bodies in finite dimensional Euclidean spaces. In particular, we refine a result of Kannan, Lovász and Simonovits on the Cheeger isoperimetric constants.

Key words and phrases. Coercive and isoperimetric inequalities, infinite dimensional spaces, concentration of measure, Markov semigroups, Gibbs measures.

Supported by EPSRC GR/N22465 and NSF GR DMS-0103929 & DMS-0405587.

Preface

During the second half of the XXth century one could observe an interesting development in the infinite dimensional analysis slowly moving it from the "experimental science" based on interesting examples to a brink of becoming something more than that. In particular during its last decade we have experienced a considerable progress in the domain of coercive inequalities and isoperimetry; as a starting point to explore the related literature we suggest [**A-B-C-F-G-M-R-S**], [**G-Z**] and [**L5**], as they contain an extensive overview and a comprehensive list of references. For the purposes of this work we would like to recall two important inequalities: the so called Logarithmic Sobolev Inequality of [**G1**] and Isoperimetric Functional Inequality introduced in [**B2**]. The first one in plain language is simply an estimate of the relative entropy in terms of the Fisher entropy. Alternatively, from the perspective of application to control the ergodicity of Markov semigroup generated by the Dirichlet operator, we could think of the relative density as given by a square of a function divided by a normalisation constant and in this setting the upper bound in question is given by the Dirichlet form as follows

$$(\mathbf{LS}_2) \qquad \operatorname{Ent}\left(|f|^2\right) \leq c \int \sum_i |\nabla_i f|^2 \, d\mu.$$

The Isoperimetric Functional Inequality is a bound on the value of an isoperimetric function $\mathcal{I}$ computed at the point equal to the expectation of a positive bounded by 1 function f by an expectation of the length of a vector which first component equals to the composition of $\mathcal{I}$ with f and the other components are given by the components of the gradient of f scaled by a constant

$$(\mathrm{IFI}_2) \qquad \mathcal{I}\left(\mu f\right) \leq \int \sqrt{\mathcal{I}\left(f\right)^2 + c \sum_i |\nabla_i f|^2} \, d\mu.$$

By an appropriate approximation of a characteristic function the right hand side converges to the surface measure of the set whereas the limit of the left hand side is simply equal to the value of the isoperimetric function at the volume of the set in question; this provides justification for the name of the inequality. The first key point is that both of these inequalities have tensorisation property, (i.e. if they hold in two measure spaces, then the product of the measures also satisfies them). The second important common point is that both of them involve the same l_2-norm on the tangent space. Naturally given a variety of probability measures one may like to ask for an optimal information; that is for inequalities reflecting possibly precisely properties of a given measure. It is certainly well known that in finite dimensions the isoperimetric functions may be different for different families of the measures. In our work we would like to emphasise the role of the metric on the tangent space showing in particular that nonequivalent in infinite dimensions l_q-metrics are associated to qualitatively different families of measures.

In brief the results and organisation of the paper is as follows. The first part is devoted to the study of the following relative entropy bounds

$$(\mathrm{LS}_q) \qquad \mathrm{Ent}\left(|f|^q\right) \leq c \int \sum_i |\nabla_i f|^q \, d\mu.$$

with $q \in (1,2)$. In Section 1, after a brief introduction, we indicate that LS_q is intimately related to a faster decay of the tails of the corresponding distributions and on an abstract level via Herbst arguments one can obtain a suitable exponential bounds. Section 2 describes the q-Poincaré inequality- the first abstract consequence of our coercive inequality- as well as a certain improvement on the lower bound of the spectral gap in finite dimension. In Section 3 we show that similarly as in [**B-G1**] for LS_2 also when in the extended region of the parameter q one can think of LS_q as a Poincaré inequality with respect to a suitable Orlicz space. This section contains also an analog of the so called Rothaus lemma (on improvements of an inequality involving additionally L_q norm of the function on its right hand side provided q-Poincaré is known). Sections 4 and 5 specialise to the probability measures on the real line providing essentially optimal bounds for the constants in the q-Poincaré and q-Log Sobolev inequality. The ideas involved here are based on utillisation of Hardy inequalities and the Orlicz norms. These results naturally extend the ones of [**B-G1**]. An easy application of these results shows that for example measures with density $(\exp -x^p)/Z$, with $p \in (2,\infty)$ being the Hölder dual to q, satisfy LS_q. Hence by the product property we recover the similar result of [**B-L**] which proof was essentially based on the Prekopa-Leindler inequality involving strong concavity properties of the log of the density. Additionally, in Section 5, we provide a class of examples of measures on the real line with smooth densities for which LS_q is true although the conditions of [**B-L**] are infinitely violated; this class of examples goes in the spirit of [**G-R**], ([**A-B-C-F-G-M-R-S**]). We come back to exponential bounds in Section 6 providing its more optimal version, generalising the one of [**B-G1**] and allowing unbounded gradients of the random variables. In this section we indicate also that such bounds together with the Rothaus lemma can be used to show LS_q in finite dimensions for measures with highly singular densities. Section 7 contains a proof of the LS_q for Gibbs measures with super-Gaussian tails, (that is such that distribution of the natural coordinate functions decay faster than Gaussian's), associated to a local specification (i.e. an a'priori given family of regular conditional expectations with respect to complements of some finite dimensional sigma algebras) which satisfies a strong mixing condition. This result is obtained by a suitable adaptation of the technology developed in the past for LS_2 and nicely confirms the robustness of the original ideas. Naturally one can be interested what, if any, consequences one can derive from LS_q for some family of Markov semigroups. Some preliminary results on this subject the reader can find in Section 8. In particular, confining to arbitrary but finite dimensions, we show that the semigroup generated by the ordinary Dirichlet operator is ultracontractive. Additionally we present there also some ergodicity estimates for the semigroups generated by the nonlinear Laplacians, (in particular obtaining an algebraic entropy decay which in the limit $q \to 2$ recovers the well known exponential decay of entropy associated with the LS_2). Sections 9 - 13 concern the high dimensional asymptotics of the isoperimetric problem for convex bodies with the uniform measure and its dependence on the underlying metric. The key result provides a lower bound for so called Sobolev constant in dependence on the intrinsic characteristic of the convex body involving inner metric and the dimension of the space. The proof utilises a great localisation technique of [**K-L-S**] which for convex bodies allows to reduce the problem to a suitable one dimensional problem. As a corollary of the estimates, by employing arguments based on the coarea formula, in Section 14 one obtains the Classical Sobolev inequality. This precise asymptotics together with some appropriate limiting arguments allow us to recover in Section 15 the LS_q inequality. Thus in the sense we have proved the equivalence of the LS_q and the infinite dimensional asymptotics of the isoperimetric problem. Finally in Section 16 we consider the following functional isoperimetric inequality IFI_q introduced in [**Z1**]

$$(\mathrm{IFI}_2) \qquad \mathcal{U}(\mu f) \leq \int \sqrt[q]{\mathcal{U}(f)^q + c\sum_i |\nabla_i f|^q}\, d\mu.$$

As shown in [**Z1**] it has a product property. Moreover, by a suitable choice of the isoperimetric function $\mathcal{U}$, in natural way it reflects known isoperimetric properties of the finite dimensional restrictions of the measures with super Gaussian tails. In this section we show the implication $IFI_q \Longrightarrow LS_q$ generalizing the known case $q = 2$. (We remark that in the case $q = 2$ the converse inequality was also proved in [**Fo**]; though at this point of time the semigroup technique does not seem to admit the necessary generalisation.)

These are the results, but at this point it may be proper to indicate where all that could be going how that could be useful. Certainly further deeper understanding the relation between the metric properties and isoperitmetry as well as coercive and functional isoperimetric inequalities would be useful and interesting. While in the above we have mentioned only the super Gaussian tails, we should mention that there exist also some results for the measures with sub Gaussian tails (that is with decay between exponential and the Gaussian). In particular an interesting inequality which has a product property was proved in [**L-O**] providing a bound for a difference between the square of L_2 and the square of L_r norm (for any $r \in [1, 2)$) in terms of the Dirichlet form. As showed by the authors such bound allows to recover the correct exponential bounds; (for an earlier work involving generalized Poincaré inequalities in some other context see [**Be**] and for the product property in a more general setting see [**L3**], [**Lie**]). Later, it was demonstrated in the epilog to [**G-Z**] that their inequality extend to a large class of Gibbs measures with sub Gaussian tails. Alternative results involving the same class of measures and bounds of entropy by higher order differential expressions were proven in [**Z2**] following natural link with classical Sobolev inequalities for probability measures of [**Ros**]. Understanding properly all interconnections as well as the isoperimetry in this case remains still a challegenge. One should hope that plentiful of interesting things may follow this research in terms of understanding the new classes of infinite dimensional equations involving nonlinear as well as differential operators of higher order.

CHAPTER 1

Introduction and notations

Let (Ω, μ) denote a probability space, and assume there is an operator Γ defined on some algebra $\mathcal{A}$ of measurable functions on Ω with the following properties:

1) for any $f \in \mathcal{A}$, $\Gamma(f)$ is a non-negative measurable function on Ω;

2) for all $f \in \mathcal{A}$ and smooth functions u on $\mathbf{R}$, $u(f) \in \mathcal{A}$ and $\Gamma(u(f)) = |u'(f)|\,\Gamma(f)$.

Introduce the entropy functional:

$$\operatorname{Ent}(g) = \int g \log g \, d\mu - \int g \, d\mu \, \log \int g \, d\mu.$$

It is well defined for all measurable $g \geq 0$ and then it is finite if and only if so is the integral $\int g \log(1+g)\, d\mu$.

One says that (Ω, μ, Γ) satisfies a logarithmic Sobolev inequality, for short LSI with constant $c \geq 0$ if, for all $f \in \mathcal{A}$,

$$\operatorname{Ent}(|f|^2) \leq c \int \Gamma(f)^2 \, d\mu. \tag{1.0.1}$$

The definition is similar to the one considered in [A-M-S].

For example, when Ω is a metric space with metric d, the "modulus of the gradient" comes naturally via the identity

$$\Gamma(f)\,(x) = |\nabla f(x)| = \limsup_{d(x,y)\to 0+} \frac{|f(x) - f(y)|}{d(x,y)}$$

with the convention that $\Gamma(f)\,(x) = 0$ at isolated points x in Ω. In this case, we may define Γ on the class $\mathcal{A}$ of all Lipschitz functions f, i.e., such that $\|f\|_{\mathrm{Lip}} < \infty$, or for a larger class of all locally Lipschitz functions (i.e., Lipschitz in a neighbourhood of any point).

Moreover, in case of the Euclidean space $\Omega = \mathbf{R}^n$ with the usual Euclidean metric $d(x,y) = |x-y|$, we clearly have $\Gamma(f)(x) = |\nabla f(x)|$ at each point $x \in \mathbf{R}^n$ where f is differentiable and has gradient $\nabla f(x)$. Since locally Lipschitz functions are differentiable almost everywhere (with respect to Lebesgue measure), we then arrive in (1.1) at the usual definition of logarithmic Sobolev inequalities.

More generally, given a number $p \in (1, +\infty]$, we may equip $\mathbf{R}^n$ with ℓ^p-metric $d(x,y) = \|x-y\|_p = \left(\sum_{i=1}^n |x_i - y_i|^p\right)^{1/p}$, and then we obtain another important modulus of gradient,

$$\Gamma(f)(x) = |\nabla f(x)|_q = \left(\sum_{i=1}^n \left|\frac{\partial}{\partial x_i} f(x)\right|^q\right)^{1/q},$$

[0]Received by the editor October 1, 2002.

where $q = \frac{p}{p-1}$ is conjugate to p.

Actually, already the one-dimensional case contains a number of open problems. One of questions of interest is to determine whether or not, a given probability measure μ satisfies LSI with some finite c. In the case of the real line $\Omega = \mathbf{R}$, the answer is known [B-G1], and we recall the obtained characterization.

Let $F(x) = \mu((-\infty, x])$, $x \in \mathbf{R}$, denote the distribution function of μ, and let p be the density of the absolutely continuous part of μ with respect to Lebesgue measure. Let m denote a median of μ. Set

$$D_0 = \sup_{x<m} \left(F(x) \log \frac{1}{F(x)}\right) \int_x^m \frac{1}{p(t)}\,dt,$$

$$D_1 = \sup_{x>m} \left((1-F(x)) \log \frac{1}{1-F(x)}\right) \int_m^x \frac{1}{p(t)}\,dt,$$

defining D_0 and D_1 to be zero in case $\mu((-\infty, m)) = 0$ or $\mu((m, +\infty)) = 0$, respectively.

Theorem 1.1. *The measure μ satisfies the log-Sobolev inequality* (1.1) *with some constant c if and only if $D_0 + D_1 < +\infty$. In this case, the optimal value of c satisfies*

$$K_0(D_0 + D_1) \le c \le K_1(D_0 + D_1),$$

where K_0 and K_1 are certain absolute positive constants.

Due to the tensorization property, the log-Sobolev inequality (1.1) can be extended to product spaces without any loss in constant. In high-dimensional spaces it can be used to recover various concentration phenomena, dependently on which classes $\mathcal{A}$ and gradients Γ are involved in (1.1). On the other hand, it is a powerful tool in the study of hypercontractivity of semi-groups (cf. [G1], [G2]). However, more specific and delicate properties of certain measures suggest to consider somewhat different forms of log-Sobolev inequalities. In particular, inequalities

$$\text{Ent}\,(\,|f|^q) \le c \int \Gamma(f)^q\,d\mu, \tag{1.0.2}$$

with (necessarily) some $q \in [1, 2]$, may serve as a certain sharpening of (1.1), describing more accurately the behaviour of distributions of Lipschitz functions. For example, as shown by O. Rothaus [R], the particular case $q = 1$ reduces (1.2) on a Riemannian manifold to a certain isoperimetric inequality.

Definition 1.2. *For short, in what follows* (1.2) *is called* LS_q*-inequality and if a measure μ satisfies it with a given coefficient $c \in (0, \infty)$ we denote that by $\mu \in \mathrm{LS}_q(c)$.*

Let us recall an argument which relates LS_q-inequality to deviations of "Lipschitz" functions from their means and therefore to the concentration of measure. The idea goes back to an unpublished letter by I. Herbst (of mid 70s), and only in late 80s it was reanimated and developed in the works [A-M-S] and [L2,3]). Let $1 < q \le 2$ and assume g is a bounded function in $\mathcal{A}$ such that $\int g\,d\mu = 0$ and $\Gamma(g) \le 1$ μ-a.e. We may apply (1.2) to the function $f = e^{\lambda g/q}$, $\lambda \ge 0$: due to the axiom 2), it belongs to $\mathcal{A}$ and, moreover, $\Gamma(f)^q \le (\frac{\lambda}{q})^q\, e^{\lambda g}$. Write the Laplace transform on g as $\int e^{\lambda g}\,d\mu = e^{\lambda v(\lambda)}$ for some function v which is clearly smooth in $\lambda > 0$

and satisfies $v(0+) = 0$ (since g has μ-mean zero). Thus, the right hand side of (1.2) does not exceed $c\,(\frac{\lambda}{q})^q\, e^{\lambda v(\lambda)}$. On the other hand, $\mathrm{Ent}\,(e^{\lambda g}) = \lambda^2 v'(\lambda) \int e^{\lambda g}\, d\mu$, so $\lambda^2 v'(\lambda) \leq c\,(\frac{\lambda}{q})^q$ and

$$v(\lambda) = \int_0^\lambda v'(t)\, dt \leq \frac{c}{q^q} \int_0^\lambda t^{q-2}\, dt = \frac{c}{q^q(q-1)}\, \lambda^{q-1}.$$

Thus, we have obtained the following bound on the Laplace transform:

$$\int e^{\lambda g}\, d\mu \leq \exp\left\{\frac{c}{q^q(q-1)}\, \lambda^q\right\}, \quad \lambda > 0. \tag{1.0.3}$$

Now, given a number $h > 0$, by Chebyshev's inequality and (1.3), we get

$$\mu\{g \geq h\} \leq e^{-\lambda h} \int e^{\lambda g}\, d\mu \leq \exp\left\{\lambda h - \frac{c}{q^q(q-1)}\, \lambda^q\right\}, \quad \lambda > 0.$$

Optimization over all λ leads to the one-sided estimate $\mu\{g \geq h\} \leq \exp\{-\frac{(q-1)^p}{c^{p-1}}\, h^p\}$, where p is conjugate to q. Applying it to $-g$, we obtain a similar estimate. Hence, we arrive at the following important conclusion:

Theorem 1.3. *Under* LS_q, $1 < q \leq 2$, *for every bounded function* $g \in \mathcal{A}$ *with mean* $a = \int g\, d\mu$, *such that* $\Gamma(g) \leq 1$ μ*-a.e.,*

$$\mu\{|g - a| \geq h\} \leq 2\exp\left\{-\frac{(q-1)^p}{c^{p-1}}\, h^p\right\}, \quad h > 0, \tag{1.0.4}$$

where $p = \frac{q}{q-1}$.

The main feature of the inequality (1.4) is that, in case $q < 2$, it indicates a much stronger decay of the tails of g in comparison with the classical "Gaussian" case $q = 2$. Of course, in the reasonable situation such as an abstract probability metric space, the assumption on boundedness of g can be removed (via usual approximations). Thus, (1.4) remains to hold for unbounded Lipschitz g, as well. Moreover, application of (1.4) to functions of the form $\mathrm{dist}(A, x)$ leads to certain concentration inequalities often complementing or illustrating the so-called concentration of measure phenomenon (cf. e.g. [L4,5]). As for (1.3), one may wonder whether or not it is possible to replace the Lipschitz condition by a more flexible property of the modulus of gradient; we will concern this question in section 6.

LS_q inequalities (1.2), together with deviation inequality (1.4), are known to hold for probability measures on $\mathbf{R}^n$ with sufficiently "convex" densities, cf. [B-L]. Here we will mainly specialize on the question of characterization of μ on the real line $\mathbf{R}$ which thus satisfy the inequality

$$\mathrm{Ent}\,(|f|^q) \leq c \int_{-\infty}^{+\infty} |f'(x)|^q\, d\mu(x)$$

in the class of all smooth functions f, or equivalently, locally Lipschitz f, up to some constant c.

The proof of Theorem 1.1 given in [B-G1] uses a particular L_2-case of result of M. Artola, G. Talenti and G. Tomaselli on Hardy-type inequalities with weights on the real line (cf. [Mu], [**M2**]). Here we have to use it in the full volume involving L_q spaces. A non-trivial point is that the general case requires a more careful analysis of possible behaviour of the entropy functional $\mathrm{Ent}\,(|f|^q)$ and of what is connected with it,

$$\mathcal{L}_q(f) = \sup_{a \in \mathbf{R}} \mathrm{Ent}\,(|f+a|^q). \tag{1.0.5}$$

Indeed, the log-Sobolev inequality (1.1) may formally be strengthened as

$$\mathcal{L}_q(f) \leq c \int \Gamma(f)^q \, d\mu.$$

Note that, when f is bounded, $\mathrm{Ent}\,(|f+a|^q)$ is of order $a^{q-2}\,\mathrm{Var}(f)$, for large values of a (where $\mathrm{Var}(f)$ denotes variance of f with respect to μ). So, in case $q > 2$, $\mathcal{L}_q(f) = +\infty$ and thus (1.2) can never hold. In the case $1 \leq q < 2$, we have $\mathrm{Ent}\,(|f+a|^q) \to 0$, as $|a| \to +\infty$, and so we cannot derive from (1.2) a Poincaré-type inequality similarly to the case $q = 2$. This is a typical example illustrating some features of these more general inequalities.

CHAPTER 2

Poincaré-type inequalities

Here, in the setting of a probability metric space (Ω, d, μ), we deduce from (1.2) a corresponding Poincaré-type inequality:

Theorem 2.1. *Under* LS_q*-inequality with constant* c*, for any locally Lipschitz function* f *on* Ω*, we have the following Poincaré type inequality called later on* SG_q-inequality

$$\int \left| f - \int f \, d\mu \right|^q d\mu \leq \frac{4c}{\log 2} \int |\nabla f|^q \, d\mu. \tag{2.0.1}$$

More precisely, as soon as $|\nabla f| \in L_q(\mu)$*, we have* $f \in L_q(\mu)$*, and* (2.1) *holds true.*

Recall that $|\nabla f(x)| = \limsup_{d(x,y)\to 0^+} \frac{|f(x)-f(y)|}{d(x,y)}$. In particular, this definition is applied on the real line, so f does not need to be differentiable everywhere (correspondingly, we understand partial derivatives $\frac{\partial f(x,y)}{\partial y}$ for functions defined on the product space $\Omega \times \mathbf{R}$).

For the proof of (2.1), we need a general

Lemma 2.2. *For any non-negative measurable function* g *on* Ω*,*

$$\mathrm{Ent}\,(g) \geq -\log \mu\{g > 0\} \int g \, d\mu. \tag{2.0.2}$$

Proof. The distribution of g under μ represents a probability measure on the positive half-axis $[0, +\infty)$ and may be written as $(1-\alpha)\delta_0 + \alpha\nu$ where $\alpha = \mu\{g > 0\}$ and where ν is concentrated on $(0, +\infty)$. In terms of a random variable, say ξ, having the distribution ν, we thus may write

$$\int g \, d\mu = \int_0^{+\infty} x \, d\left((1-\alpha)\delta_0 + \alpha\nu\right) = \alpha \mathbf{E}\,\xi,$$

$$\int g \log g \, d\mu = \int_0^{+\infty} x \log x \, d\left((1-\alpha)\delta_0 + \alpha\nu\right) = \alpha \mathbf{E}\,\xi \log \xi,$$

so,

$$\begin{aligned} \mathrm{Ent}\,(g) &= \alpha\,\mathbf{E}\,\xi \log \xi - (\alpha \mathbf{E}\,\xi) \log(\alpha \mathbf{E}\,\xi) \\ &= \alpha\,\mathrm{Ent}\,(\xi) - \alpha \log \alpha\,\mathbf{E}\,\xi \;\geq\; -\log \alpha\;(\alpha \mathbf{E}\,\xi). \end{aligned}$$

Proof of Theorem 2.1. Let f be a locally Lipschitz function from $L_q(\Omega, \mu)$. Without loss of generality, assume that zero is a median of f, i.e., $\mu\{f > 0\} \leq \frac{1}{2}$ and $\mu\{f < 0\} \leq \frac{1}{2}$. First assume that $\mu\{f = 0\} = 0$.

Set $f^+ = \max(f, 0)$, $f^- = \max(-f, 0)$. These functions are also locally Lipschitz, and we can apply to them the log-Sobolev inequality (1.2). The moduli of gradients of f^+ and f^- are respectively vanishing on the open sets $\{f < 0\}$, $\{f > 0\}$ and coincide with $|\nabla f|$ on $\{f > 0\}$, $\{f < 0\}$. Since $\mu\{f = 0\} = 0$, the further application of (2.2) to $g = (f^+)^q$ and $g = (f^-)^q$ gives respectively

$$\begin{aligned} c \int_{\{f>0\}} |\nabla f|^q \, d\mu &\geq \operatorname{Ent}\left((f^+)^q\right) \geq \log 2 \int_{\{f>0\}} (f^+)^q \, d\mu, \\ c \int_{\{f<0\}} |\nabla f|^q \, d\mu &\geq \operatorname{Ent}\left((f^-)^q\right) \geq \log 2 \int_{\{f<0\}} (f^-)^q \, d\mu. \end{aligned}$$

Summing these yields

$$c \int |\nabla f|^q \, d\mu \geq \log 2 \int |f|^q \, d\mu = \log 2 \, \|f\|_q^q.$$

But $\|f - \mathbf{E}\, f\|_q \leq \|f\|_q + \|f\|_1 \leq 2\|f\|_q$, so $\|f - \mathbf{E}\, f\|_q^q \leq 2^q \|f\|_q^q \leq 4\|f\|_q^q$, and this is what we need.

Note that, with a similar argument, we could proceed without the assumption $\mu\{f = 0\} = 0$ and gain the double constant in (2.1) by using the estimates $|\nabla f^+| \leq |\nabla f|$, $|\nabla f^-| \leq |\nabla f|$ which hold true on the whole space Ω. There is another simple way to avoid this assumption which however saves the same constant.

Namely, let ξ be a random variable with mean zero and whose distribution, say, ν satisfies on the real line LS_q (1.2) with constant c_q. For example, the measure ν may have density of the form $\mathrm{const}\, e^{-|x|^p}$ where p is conjugate to q. Hence, the distribution ν_ε of the random variable $\varepsilon\xi$ satisfies LS_q with constant $c_q \varepsilon^q$. By the tensorization property of entropy, the log-Sobolev inequality (1.2) implies

$$\begin{aligned} \operatorname{Ent}_{\mu\otimes\nu_\varepsilon}(|g(x,y)|^q) &\leq c \int_R \int_\Omega |\nabla_x g(x,y)|^q \, d\mu(x) d\nu_\varepsilon(y) + \\ &\quad c_q \varepsilon^q \int_R \int_\Omega |\partial_y g(x,y)|^q \, d\mu(x) d\nu_\varepsilon(y) \end{aligned}$$

where g is now locally Lipschitz on the product space $\Omega \times \mathbf{R}$ equipped with the product measure $\mu \otimes \nu_\varepsilon$. Thus, the product probability space $(\Omega \times \mathbf{R}, \mu \otimes \nu_\varepsilon)$ satisfies LS_q with respect to the (non-canonical) gradient functional

$$\Gamma_\varepsilon g(x,y) = \left(c\,|\nabla_x g(x,y)|^q + c_q \varepsilon^q \, |\partial_y g(x,y)|^q\right)^{1/q}.$$

Now, given a locally Lipschitz function f from $L_q(\Omega, \mu)$, the function $g(x,y) = f(x) + y$ is also locally Lipschitz, belongs to $L_q(\Omega \times \mathbf{R}, \mu \otimes \nu_\varepsilon)$, and is equidistributed, under $\mu \otimes \nu_\varepsilon$, with $f + \varepsilon\xi$ where ξ is viewed as being independent of f. In particular, g has a smooth positive density on the whole line, so the median m of g is uniquely determined and, moreover, $\mu \otimes \nu_\varepsilon\{g = m\} = 0$. The functional Γ_ε enjoys many properties of the modulus of gradient, at least in part concerning the function g. Namely, $\Gamma_\varepsilon g^+ = \Gamma_\varepsilon g$ on $\{g > 0\}$, $\Gamma_\varepsilon g^+ = 0$ on $\{g < 0\}$, and similarly for g^-. Therefore, one can repeat the previous argument and thus arrive, according to (2.1) and the first step, at

$$\begin{aligned} \mathbf{E} \int \left| (f(x) + \varepsilon\xi) - \int f \, d\mu \right|^q d\mu(x) &\leq \frac{4}{\log 2} \int (\Gamma_\varepsilon g)^q \, d\mu \otimes \nu_\varepsilon \\ &= \frac{4}{\log 2} \int_R \int_\Omega \left(c\,|\nabla_x f(x)|^q + c_q \varepsilon^q\right) d\mu(x) d\nu_\varepsilon(y) \\ &= \frac{4}{\log 2} \left(c \int_\Omega |\nabla f|^q \, d\mu + c_q \varepsilon^q \right). \end{aligned}$$

It remains to let $\varepsilon \to 0$. Proposition 2.1 is now proved.

It will also be useful to note that any Poincaré-type inequality in L^q space with $q \in [1, 2)$,

$$\int \left| f - \int f \, d\mu \right|^q d\mu \leq C_q \int |\nabla f|^q \, d\mu, \tag{2.0.3}$$

holding true in the class of all locally Lipschitz functions f on Ω, is stronger than the usual spectral gap,

$$\int \left| f - \int f \, d\mu \right|^2 d\mu \leq C_2 \int |\nabla f|^2 \, d\mu. \tag{2.0.4}$$

Namely, we have:

Proposition 2.3. *For any $q \in [1, 2]$, the optimal constants in* (2.3) - (2.4) *are connected by*

$$C_2^{1/2} \leq 6\, C_q^{1/q}.$$

Remark 2.4. Since we are still dealing here with the abstract probability metric space (Ω, d, μ), the modulus of gradient has unique sense and should not be mixed in case, for example, $\Omega = \mathbf{R}^n$ or $\Omega = \mathbf{R}^\infty$ with other natural gradients such as $\|\nabla f\|_q$ in (2.3).

Proof of Proposition 2.3. Let $1 \leq q < 2$. First we transform (2.3) to an analogue with the median in the place of the mean. Let f be a locally Lipschitz function in $L^q(\mu)$ such that its μ-mean is zero. Thus, by (2.3),

$$\int |f|^q \, d\mu \leq C_q \int |\nabla f|^q \, d\mu. \tag{2.0.5}$$

Denote by $m = m(f)$ its median so that $\mu\{f \leq m\} \geq \frac{1}{2}$ and $\mu\{f \geq m\} \geq \frac{1}{2}$. Assume for definiteness that $m > 0$. Since by Chebyshev's inequality $\mu\{f \geq m\} \leq \|f\|_q^q / m^q$, we may conclude that $m^q \leq 2\, \|f\|_q^q$. Thus, in general,

$$|m| \leq 2^{1/q}\, \|f\|_q.$$

Hence, $\|f - m\|_q \leq \left(1 + 2^{1/q}\right) \|f\|_q$, and by (2.5),

$$\int |f - m(f)|^q \, d\mu \leq \left(1 + 2^{1/q}\right)^q C_q \int |\nabla f|^q \, d\mu. \tag{2.0.6}$$

This inequality is translation invariant, so it holds true for all locally Lipschitz functions in $L^q(\mu)$. Now, take such a function and assume moreover that f is square integrable, and that $m(f) = 0$. Assume that $\mu\{f = 0\} = 0$ and consider the locally Lipschitz functions $f^+ = \max(f, 0)$, $f^- = \max(-f, 0)$. As in the proof of Proposition 2.1, let us note once more that the moduli of gradients of f^+ and f^- are respectively vanishing on the open sets $\{f < 0\}$, $\{f > 0\}$ and coincide with $|\nabla f|$ on $\{f > 0\}$, $\{f < 0\}$. Since $m(f^+) = m(f^-) = 0$, the application of (2.6) to $(f^+)^{2/q}$ and $(f^-)^{2/q}$ gives respectively

$$\begin{aligned}
\int_{\{f>0\}} |f|^2 \, d\mu &\leq \left(1 + 2^{1/q}\right)^q C_q \left(\frac{2}{q}\right)^q \int_{\{f>0\}} |f|^{2-q}\, |\nabla f|^q \, d\mu, \\
\int_{\{f<0\}} |f|^2 \, d\mu &\leq \left(1 + 2^{1/q}\right)^q C_q \left(\frac{2}{q}\right)^q \int_{\{f<0\}} |f|^{2-q}\, |\nabla f|^q \, d\mu.
\end{aligned}$$

Summing these yields

$$\int |f|^2 \, d\mu \ \le \ C_q \left(\frac{2\,(1+2^{1/q})}{q}\right)^q \int |f|^{2-q}\, |\nabla f|^q \, d\mu. \tag{2.0.7}$$

Since the conjugate of $\frac{2}{q}$ is $\frac{2}{2-q}$, the latter integral can be estimated with the help of Hölder's inequality by $\|\,|f|^{2-q}\,\|_{2/(2-q)}\,\|\,|\nabla f|^q\,\|_{2/q} = \|f\|_2^{2-q}\,\|\nabla f\|_2^q$. Hence, (2.7) implies

$$\int |f|^2 \, d\mu \ \le \ C_q^{2/q} \left(\frac{2\,(1+2^{1/q})}{q}\right)^2 \int |\nabla f|^2 \, d\mu.$$

Since $\int \left|f - \int f\, d\mu\right|^2 \, d\mu \le \int |f|^2 \, d\mu$, we arrive at

$$\int \left|f - \int f\, d\mu\right|^2 \, d\mu \ \le \ C_q^{2/q} \left(\frac{2\,(1+2^{1/q})}{q}\right)^2 \int |\nabla f|^2 \, d\mu$$

where the assumption that the function f has median zero may already be omitted. The assumption that $\mu\{f = m(f)\} = 0$ can be omitted by an argument used in the proof of Proposition 2.1. Thus, we arrived at the claim (2.4) with constant

$$C_2^{1/2} \le \frac{2\,(1+2^{1/q})}{q}\, C_q^{1/q} \le 6\, C_q^{1/q}.$$

Combining Propositions 2.1 and 2.3, we may conclude that LS_q with constant c implies the usual Poincaré inequality with constant $C_2 \le 36\,(\frac{4c}{\log 2})^{2/q}$. This spectral gap can actually be improved. Indeed, when assuming in the proof of Proposition 2.1 that f has median zero, we derived a Poincaré-type inequality (2.1) with a better constant,

$$\int |f|^q \, d\mu \le \frac{c}{\log 2} \int |\nabla f|^q \, d\mu.$$

And under the same assumption, in the proof of Proposition 2.3 we deduced from the latter inequality the following one:

$$\int |f|^2 \, d\mu \ \le \ \left(\frac{c}{\log 2}\right)^{2/q} \left(\frac{2}{q}\right)^2 \int |\nabla f|^2 \, d\mu.$$

The constant is maximized at $q = 1$, and this leads to

Corollary 2.5. *Under* LS_q*-inequality with constant* c*, for any locally Lipschitz function* f *on* Ω*,*

$$\int \left|f - \int f\, d\mu\right|^2 \, d\mu \le 9\, c^{2/q} \int |\nabla f|^2 \, d\mu.$$

CHAPTER 3

Entropy and Orlicz spaces

In order to further explore LS_q-inequalities, one has to bring the attention to some delicate properties, although yet general, of the entropy functional. In this section, we collect mainly those of them which connect this functional with norm in Orlicz space.

Given a Young function $N : \mathbf{R} \to [0, +\infty)$, i.e., an even, convex function with $N(0) = 0$, $N(x) > 0$ for $x > 0$, the Orlicz space $L_N = L_N(\Omega, \mu)$ consists of all measurable functions f such that

$$\|f\|_N = \inf\left\{\lambda > 0 : \int N(f/\lambda)\, d\mu \leq 1\right\} < +\infty.$$

Any Young function N strictly increases on $[0, +\infty)$ so an inverse $N^{-1} : [0, +\infty) \to [0, +\infty)$ exists. When $N(x) = |x|^q$ $(1 \leq q < +\infty)$, L_N is the usual Lebesgue space with norm $\|f\|_q$. We will consider the Orlicz norms for the two Young functions

$$N_q(x) = |x|^q \log(1 + |x|^q) \quad \text{and} \quad \Psi(x) = |x| \log(1 + |x|).$$

Eventually, we will obtain here the following characterization.

Proposition 3.1. *Let $1 \leq q \leq 2$. In the class of all locally Lipschitz integrable functions on a probability metric space (Ω, d, μ), the LS_q-inequality with constant c is equivalent to*

$$\left\|f - \int f\, d\mu\right\|_{N_q}^q \leq C \int |\nabla f|^q\, d\mu. \tag{3.0.1}$$

Moreover, the optimal constants there satisfy $\frac{1}{7}\, C \leq c \leq 16\, C$.

Thus, up to a numerical constant, LS_q can be expressed as a certain Poincaré-type inequality in the Orlicz space $L_{N_q}(\Omega, \mu)$.

We start with observations in the setting of a general probability space (Ω, μ).

Lemma 3.2. *For any function $f \in L_{N_q}(\mu)$, $q \geq 1$,*

$$\|f\|_q^q \leq \frac{5}{4}\|f\|_{N_q}^q.$$

Proof. The optimal constant is attained at $f = 1$ and thus equals $\Psi^{-1}(1) < \frac{5}{4}$.

Lemma 3.3. *For any function $f \in L_{N_q}(\mu)$, $q \geq 1$,*

$$\|f\|_{N_q}^q \leq \mathrm{Ent}\,(|f|^q) + \|f\|_q^q.$$

Proof. We may assume $\|f\|_q = 1$ so that $\mathrm{Ent}\,(|f|^q) = \int |f|^q \log |f|^q \, d\mu$. Using $x\log(1+x) \leq x \log x + 1$, valid for all $x \geq 0$, we get

$$\int N_q(f)\, d\mu \leq \mathrm{Ent}\,(|f|^q) + 1 \equiv \alpha.$$

As follows from the definition, for all $t \in [0,1]$ and $x \in \mathbf{R}$, $N_q(tx) \leq t^q N_q(x)$. Hence, applying this to $t = \frac{1}{\alpha}$, we have

$$\int N_q\left(\frac{f}{\alpha^{1/q}}\right) d\mu \leq \frac{1}{\alpha} \int N_q(f)\, d\mu \leq 1$$

which means that $\|f\|_{N_q} \leq \alpha^{1/q}$.

Lemma 3.4. *For any function* $f \in L_{N_q}(\mu)$, $q \geq 1$,

$$\mathrm{Ent}\,(|f|^q) \leq \frac{e+1}{e}\, \|f\|_{N_q}^q.$$

Proof. We may now assume that $\|f\|_{N_q} = 1$ so that $\int |f|^q \log(1+|f|^q)\, d\mu = 1$. Put $g = |f|^q$. The function $-x\log x$ attains its maximum at $x = 1/e$ equal to $1/e$. Therefore,

$$\mathrm{Ent}\,(g) \leq \int g \log g \, d\mu + \frac{1}{e} \leq \int g \log(1+g)\, d\mu + \frac{1}{e} = \frac{e+1}{e}.$$

Remark 3.5. Combining Lemma 3.2 with Lemma 3.4, and recalling Lemma 3.3, we may write a two-sided estimate

$$\frac{1}{3}\left(\mathrm{Ent}\,(|f|^q) + \|f\|_q^q\right) \leq \|f\|_{N_q}^q \leq \mathrm{Ent}\,(|f|^q) + \|f\|_q^q.$$

In the case $q = 2$, the proof of Proposition 3.1 is given in [B-G1]. One of the steps of the proof is based on an important observation of O. S. Rothaus [R] who showed, in the setting of an arbitrary probability space (Ω, μ), that

$$\mathcal{L}_2(f) = \sup_{a \in \mathbf{R}} \mathrm{Ent}\,(|f+a|)^2 \leq \mathrm{Ent}\,(f^2) + 2\int f^2\, d\mu$$

whenever $\int f d\mu = 0$. We need an appropriate generalization.

Lemma 3.6. *For any function* $f \in L_{N_q}(\Omega,\mu)$, $1 \leq q \leq 2$, *such that* $\int f\, d\mu = 0$,

$$\mathcal{L}_q(f) \leq 16\, \|f\|_{N_q}^q.$$

Recall that, according to remark 3.5, $\|f\|_{N_q}^q$ is comparable with $\mathrm{Ent}\,(|f|^q) + \|f\|_q^q$.

Proof of Lemma 3.6. By homogeneity, assume $\|f\|_{N_q} = \frac{1}{2}$.
Using Rothaus' estimate, we may write

$$\begin{aligned} \mathrm{Ent}\,(|f+a|^q) \ &\leq \ \mathrm{Ent}\left(\left(|f+a|^{\frac{q}{2}} - \int |f+a|^{\frac{q}{2}}\, d\mu\right)^2\right) \\ &+ \ 2\int \left(|f+a|^{\frac{q}{2}} - \int |f+a|^{\frac{q}{2}}\, d\mu\right)^2 d\mu. \end{aligned}$$

The expression under the integral sign in the second term can be written as

$$\left(\int \left(|f(\omega)+a|^{\frac{q}{2}} - |f(\omega')+a|^{\frac{q}{2}}\right) d\mu(\omega')\right)^2,$$

and by Cauchy-Schwarz, it is bounded by

$$\int \left(|f(\omega)+a|^{\frac{q}{2}} - |f(\omega')+a|^{\frac{q}{2}}\right)^2 d\mu(\omega').$$

Since $q \leq 2$, we may apply a simple inequality $\big|\,|x+a|^{\frac{q}{2}} - |y+a|^{\frac{q}{2}}\big|^2 \leq |x-y|^q$, valid for all real x, y and a, and estimate the above integral by $\int |f(\omega)-f(\omega')|^q \, d\mu(\omega')$. Thus, the second term does not exceed $2\,\|f(\omega)-f(\omega')\|_q^q \leq 2\cdot 2^q\,\|f\|_q^q \leq \frac{5}{2}$ (according to Lemma 3.2 and by the assumption that $\|f\|_{N_q} = \frac{1}{2}$).

A similar argument applies to the first entropy term. Using a general inequality $\mathrm{Ent}\,(g) \leq \int \Psi(g)\,d\mu + \frac{1}{e}$ and the property that Ψ is a Young function, this term can be bounded by

$$\begin{aligned}
&\int \Psi\left(\left(|f(\omega)+a|^{\frac{q}{2}} - \int |f(\omega')+a|^{\frac{q}{2}}\,d\mu(\omega')\right)^2\right) d\mu(\omega) + \frac{1}{e} \\
&\qquad \leq \int \Psi\left(\int |f(\omega)-f(\omega')|^q\,d\mu(\omega')\right) d\mu(\omega) + \frac{1}{e} \\
&\qquad \leq \int\!\!\int \Psi(|f(\omega)-f(\omega')|^q)\,d\mu(\omega)d\mu(\omega') + \frac{1}{e} \\
&\qquad = \int\!\!\int N_q(f(\omega)-f(\omega'))\,d\mu(\omega)d\mu(\omega') + \frac{1}{e}.
\end{aligned}$$

Since $\|f\|_{N_q} = \frac{1}{2}$, we have $\|f(\omega)-f(\omega')\|_{N_q} \leq \|f(\omega)\|_{N_q} + \|f(\omega')\|_{N_q} \leq 1$, so the last double integral is bounded by 1.

Combining the estimates for the both terms, we get

$$\mathrm{Ent}\,(|f+a|^q) \leq \left(1+\frac{1}{e}\right) + \frac{5}{2} = \left(\frac{1}{e} + 3.5\right)\cdot 2^q\,\|f\|_{N_q}^q.$$

The constant does not exceed 16, and this finishes the proof.

Proof of Proposition 3.1. First assume the Poincaré-type inequality (3.1) holds true with constant C. Then, by Lemma 3.6,

$$\mathrm{Ent}\,(|f|^q) \leq \mathcal{L}_q(f) \leq 16\left\|f - \int f\,d\mu\right\|_{N_q}^q \leq 16\,C\int |\nabla f|^q\,d\mu.$$

Hence, LS_q holds true with $c \leq 16\,C$. Conversely, start with LS_q with constant c and assume $\int f\,d\mu = 0$. By Proposition 2.1, $\|f\|_q^q \leq \frac{4c}{\log 2}\int |\nabla f|^q\,d\mu$. Therefore, by Lemma 3.3,

$$\|f\|_{N_q}^q \leq \mathrm{Ent}\,(|f|^q) + \|f\|_q^q \leq \left(c + \frac{4c}{\log 2}\right)\int |\nabla f|^q\,d\mu.$$

Hence, $C \leq c + \frac{4c}{\log 2} \leq 7c$. Proposition 3.1 is now proved.

CHAPTER 4

LS_q and Hardy-type inequalities on the line

When $\Omega = \mathbf{R}$, the Poincaré-type inequality (3.1) in the Orlicz space can further be reduced to a Hardy-type inequality for the Orlicz space norm $\| \cdot \|_\Psi$.

Let $m = m(\mu)$ denote a median of the probability measure μ on $\mathbf{R}$. It is any number such that $\mu\{(-\infty, m]\} \geq \frac{1}{2}$ and $\mu\{[m, +\infty)\} \geq \frac{1}{2}$.

For functions f on $\mathbf{R}$, we denote $f_0 = f\,\mathbf{1}_{(-\infty,m]}$ and $f_1 = f\,\mathbf{1}_{[m,+\infty)}$.

Proposition 4.1. *Let $1 \leq q \leq 2$. Assume that LS_q-inequality holds true on the line with constant c, i.e., for any smooth function f on $\mathbf{R}$,*

$$\mathrm{Ent}\,(|f|^q) \leq c \int_{-\infty}^{\infty} |f'(x)|^q \, d\mu(x). \tag{4.0.1}$$

Then, for any smooth function f on $\mathbf{R}$ with $f(m) = 0$,

$$\|\, |f_0|^q \,\|_\Psi + \|\, |f_1|^q \,\|_\Psi \leq d \int_{-\infty}^{\infty} |f'(x)|^q \, d\mu(x) \tag{4.0.2}$$

with $d = 42\,c$. Conversely, (4.2) implies (4.1) with $c = 144\,d$.

The inequality (4.2) as an equivalent form for logarithmic Sobolev inequalities on the real line will be used in the next section to characterize probability measures μ satisfying (4.1). The proof of Proposition 4.1 is given in this section.

Clearly, (4.1) as well (4.2) may be extended from the class of all smooth functions to the class of all absolutely continuous functions (with the same condition $f(m) = 0$ in the case of (4.2)).

To prove Proposition 4.1, we need an elementary general lemma.

Lemma 4.2. *For any function $f \in L_{N_q}(\Omega, \mu)$, $q \geq 1$,*

$$\left\| f - \int f \, d\mu \right\|_{N_q} \leq 2 \, \|f\|_{N_q}.$$

Conversely, if $1 \leq q \leq 2$ and $f = 0$ on some set $A \subset \Omega$ with $\mu(A) \geq \frac{1}{2}$, we also have

$$\|f\|_{N_q}^q \leq 6 \left\| f - \int f \, d\mu \right\|_{N_q}^q.$$

Proof. As already noted in the proof of Lemma 3.2, the optimal constant in the general inequality $\|f\|_1 \leq \alpha \, \|f\|_{N_q}$ is attained at $f = 1$, so for all f we have $\|f\|_1 \, \|\,1\,\|_{N_q} \leq \|f\|_{N_q}$. Therefore,

$$\left\| f - \int f \, d\mu \right\|_{N_q} \leq \|f\|_{N_q} + \left\| \int f \, d\mu \right\|_{N_q} = \|f\|_{N_q} + \|f\|_1 \, \|\,1\,\|_{N_q} \leq 2 \, \|f\|_{N_q}.$$

This proves the first inequality. In order to prove the second one, with some absolute constant α, we first need to derive an inequality

$$\left|\int f\,d\mu\right| \le \alpha \left\|f - \int f\,d\mu\right\|_{N_q}$$

valid in the class of all functions f in L_{N_q} vanishing on the set A. Assuming that $a = \int f\,d\mu \neq 0$, the above is equivalent to

$$\int N_q\left(\frac{\alpha\,(f-a)}{a}\right) d\mu \ge 1.$$

But on the set A, the function under the integral sign is just the constant $N_q(\alpha)$, so it suffices to pick α satisfying $N_q(\alpha) \ge 2$. Consequently, we may take $\alpha = N_q^{-1}(2) = \left(\Psi^{-1}(2)\right)^{1/q}$. Now,

$$\|f\|_{N_q} \le \left\|f - \int f\,d\mu\right\|_{N_q} + \left\|\int f\,d\mu\right\|_{N_q} \le \left(1 + \left(\frac{\Psi^{-1}(2)}{\Psi^{-1}(1)}\right)^{1/q}\right) \left\|f - \int f\,d\mu\right\|_{N_q}.$$

It remains to note that $\Psi^{-1}(1) > 1$, $\Psi^{-1}(2) < 2$, so $\left(1 + \left(\frac{\Psi^{-1}(2)}{\Psi^{-1}(1)}\right)^{1/q}\right)^q < \left(1 + 2^{1/q}\right)^q < 6$, for all $q \in [1,2]$.

Proof of Proposition 4.1. First we derive (4.2) from (4.1). By Proposition 3.1, for any smooth function f from $L_{N_q}(\mathbf{R},\mu)$,

$$\left\|f - \int f\,d\mu\right\|_{N_q}^q \le 7c \int_{-\infty}^{\infty} |f'(x)|^q\,d\mu(x).$$

Applying it to f_0 and f_1, we get

$$\left\|f_0 - \int f_0\,d\mu\right\|_{N_q}^q \le 7c \int_{(-\infty,m)} |f'(x)|^q\,d\mu(x),$$

$$\left\|f_1 - \int f_1\,d\mu\right\|_{N_q}^q \le 7c \int_{(m,+\infty)} |f'(x)|^q\,d\mu(x).$$

Applying the second inequality of Lemma 4.2 and the general identity $\|g\|_{N_q}^q = \|\,|g|^q\,\|_\Psi$, we get

$$\|\,|f_0|^q\,\|_\Psi \le 42\,c \int_{(-\infty,m)} |f'(x)|^q\,d\mu(x),$$

$$\|\,|f_1|^q\,\|_\Psi \le 42\,c \int_{(m,+\infty)} |f'(x)|^q\,d\mu(x).$$

Adding these inequalities, we obtain (4.2) with $d = 42c$.

To derive (4.1) from (4.2), first assume that $f(m) = 0$. Since $f = f_0 + f_1$,

$$\left\|f - \int f\,d\mu\right\|_{N_q} \le \left\|f_0 - \int f_0\,d\mu\right\|_{N_q} + \left\|f_1 - \int f_1\,d\mu\right\|_{N_q}.$$

On the other hand, by Lemma 3.2, $\left|\int f_0\, d\mu\right| \leq \left(\frac{5}{4}\right)^{1/q} \|f_0\|_{N_q}$ and similarly for f_1. Hence,

$$\begin{aligned}
\left\| f - \int f\, d\mu \right\|_{N_q}^q &\leq \left(1 + \left(\frac{5}{4}\right)^{1/q}\right)^q \left(\|f_0\|_{N_q} + \|f_1\|_{N_q}\right)^q \\
&\leq q\left(1 + \left(\frac{5}{4}\right)^{1/q}\right)^q \left(\|f_0\|_{N_q}^q + \|f_1\|_{N_q}^q\right) \\
&\leq 9\left(\|f_0\|_{N_q}^q + \|f_1\|_{N_q}^q\right) = 9\left(\|\,|f_0|^q\,\|_\Psi + \|\,|f_1|^q\,\|_\Psi\right).
\end{aligned}$$

By the assumption (4.2) applied to f_0 and f_1, we thus get

$$\left\| f - \int f\, d\mu \right\|_{N_q}^q \leq 9d \int_{-\infty}^{+\infty} |f'(x)|^q\, d\mu(x).$$

This inequality is invariant under translations $f \to f + \mathrm{const}$, so it holds without the condition $f(m) = 0$. At last, by Proposition 3.1, it implies that

$$\mathrm{Ent}\left(|f|^q\right) \leq 16 \cdot 9d \int_{-\infty}^{+\infty} |f'(x)|^q d\mu(x),$$

and Proposition 4.1 follows.

CHAPTER 5

Probability measures satisfying LS_q-inequalities on the real line

Using Proposition 4.1, we will give in this section a direct characterization of probability measures μ on the real line $\mathbf{R}$ satisfying the logarithmic Sobolev inequality

$$\mathrm{Ent}\,(|f|^q) \leq c \int_{-\infty}^{+\infty} |f'(x)|^q \, d\mu(x) \tag{5.0.1}$$

with some (finite) constant c for all smooth functions f on $\mathbf{R}$. This part is very similar to the case $q = 2$ studied in [B-G1]: the basic tool is the following theorem due to M. Artola, G. Talenti and G. Tomaselli (cf. [Mu]) on the optimal constant $A_q = A_q(\nu, \lambda)$ in the Hardy-type inequality with weights

$$\int_0^{+\infty} |f(x)|^q \, d\nu(x) \leq A_q \int_0^{+\infty} |f'(x)|^q \, d\lambda(x).$$

Here f is supposed to be an arbitrary smooth function on $[0, +\infty)$ such that $f(0) = 0$, and ν and λ are (non-negative) Borel measures on $[0, +\infty)$. Denote by $p_\lambda = p_\lambda(x)$ the absolutely continuous component of λ with respect to Lebesgue measure, and define the constant $B_q = B_q(\nu, \lambda)$ as follows:

$$B_q(\nu, \lambda) \;=\; \sup_{x>0} \nu([x, +\infty)) \left(\int_0^x \frac{dt}{p_\lambda(t)^{1/(q-1)}} \right)^{q-1}.$$

Theorem 5.1. ([Mu], [M2]) *For any $q \in [1, +\infty)$, $B_q \leq A_q \leq \frac{q^q}{(q-1)^{q-1}} B_q$.*

The quantity $\frac{q^q}{(q-1)^{q-1}}$ represents an increasing function in q, so, in the range of interest $1 \leq q \leq 2$, it is bounded by 4.

Theorem 5.1 has the following natural generalization. Consider a Borel measure ν on $[0, +\infty)$ and a Banach space $(X, \|\cdot\|)$ of Borel measurable functions on $[0, +\infty)$ (with usual factorization with respect to measure ν) such that

1) $f \leq |g|$ ν-a.e., $g \in X$ implies $f \in X$ and $\|f\| \leq \|g\|$, for all Borel measurable functions f;

2) any pointwise non-decreasing sequence f_n of non-negative functions in X converging pointwise to a function $f \in X$ satisfies $\|f_n\| \to \|f\|$.

By property 1), X is an ideal Banach space, and property 2) is called order semi continuity of the norm (cf. [K-A]). For an ideal Banach space X, the last property is equivalent to a representation of the norm in X in the form

$$\|f\| = \sup_{g \in \mathcal{G}} \int_0^{+\infty} |f(x)| g(x)\, d\nu(x), \tag{5.0.2}$$

for some family $\mathcal{G}$ of non-negative Borel measurable functions g on $[0, +\infty)$. This statement holds in the setting of an abstract probability space (Ω, ν) ([K-A], p.190).

For these Banach spaces X, one immediately obtains by Theorem 5.1:

Corollary 5.2. *Let $A_q = A_q(X, \lambda)$ be the optimal constant in the inequality*

$$\|\, |f|^q \,\| \leq A_q \int_0^{+\infty} |f'(x)|^q \, d\lambda(x), \tag{5.0.3}$$

where $f \in X$ is an arbitrary smooth function such that $f(0) = 0$. Then, $B_q \leq A_q \leq \frac{q^q}{(q-1)^{q-1}} B_q$, where

$$B_q = B_q(X, \lambda) \;=\; \sup_{x>0} \|\mathbf{1}_{[x,+\infty)}\| \left(\int_0^x \frac{dt}{p_\lambda(t)^{1/(q-1)}} \right)^{q-1}.$$

Indeed, the measures $\nu_g(dx) = g(x)\nu(dx)$ satisfy $B_q(\nu_g, \lambda) \leq A_q(\nu_g, \lambda) \leq 4B_q(\nu_g, \lambda)$. Using the definitions (5.2) and (5.3), we get

$$A_q(X, \lambda) = \sup_{g \in \mathcal{G}} A_q(\nu_g, \lambda), \quad B_q(X, \lambda) = \sup_{g \in \mathcal{G}} B_q(\nu_g, \lambda),$$

hence, $B_q(X, \lambda) \leq A_q(X, \lambda) \leq \frac{q^q}{(q-1)^{q-1}} B_q(X, \lambda)$.

In particular, one may apply Corollary 5.2 to the Orlicz space $X = L_\Psi(\nu)$ which of course satisfies the properties 1) and 2) above. Recall that $\Psi(x) = |x| \log(1 + |x|)$. For indicator functions, we get by definition of the Orlicz norm

$$\|\mathbf{1}_{[x,+\infty)}\|_\Psi = \frac{1}{\Psi^{-1}\left(\frac{1}{\nu([x,+\infty))}\right)},$$

where Ψ^{-1} denotes the inverse function. Consequently, the optimal constant A_q in (5.3) for the norm $\|\cdot\| = \|\cdot\|_\Psi$ and for the range $1 \leq q \leq 2$ can be estimated as follows:

$$\sup_{x>0} \frac{1}{\Psi^{-1}\left(\frac{1}{\nu([x,+\infty))}\right)} \left(\int_0^x \frac{dt}{p_\lambda(t)^{1/(q-1)}} \right)^{q-1} \leq A_q$$

$$\leq 4 \sup_{x>0} \frac{1}{\Psi^{-1}\left(\frac{1}{\nu([x,+\infty))}\right)} \left(\int_0^x \frac{dt}{p_\lambda(t)^{1/(q-1)}} \right)^{q-1}.$$

In case $\nu((0, +\infty)) \leq \frac{1}{2}$, the marginal sides of these inequalities can be simplified by noting that, for any $x > 0$, the number $t = \frac{1}{\nu([x,+\infty))}$ satisfies $t \geq \frac{1}{2}$. But in this range, we have $\frac{t}{2 \log t} \leq \Psi^{-1}(t) \leq \frac{2t}{\log t}$, and so

$$\frac{1}{2} \sup_{x>0} \nu([x, +\infty)) \log \frac{1}{\nu([x, +\infty))} \left(\int_0^x \frac{dt}{p_\lambda(t)^{1/(q-1)}} \right)^{q-1} \leq A_q$$

$$8 \sup_{x>0} \nu([x, +\infty)) \log \frac{1}{\nu([x, +\infty))} \left(\int_0^x \frac{dt}{p_\lambda(t)^{1/(q-1)}} \right)^{q-1} \tag{5.0.4}$$

Using Proposition 4.1, we are ready to formulate and to prove the main result of this section.

Let μ be a Borel probability measure on $\mathbf{R}$ with distribution function $F(x) = \mu((-\infty, x])$, and density function $p = p(x)$, $x \in \mathbf{R}$, for its absolutely continuous part with respect to Lebesgue measure. Denote by m a median of μ. Define

$$\begin{aligned} D_0(q) &= \sup_{x<m} F(x) \log \frac{1}{F(x)} \left(\int_x^m \frac{dt}{p(t)^{1/(q-1)}} \right)^{q-1}, \\ D_1(q) &= \sup_{x>m} (1 - F(x)) \log \frac{1}{1 - F(x)} \left(\int_m^x \frac{dt}{p(t)^{1/(q-1)}} \right)^{q-1}. \end{aligned}$$

defining $D_0(q)$ and $D_1(q)$ to be zero in case $\mu((-\infty, m)) = 0$ or $\mu((m, +\infty)) = 0$, respectively.

Theorem 5.3. *Let $1 \le q \le 2$. For some positive absolute constants K_0 and K_1, the optimal value of c in the logarithmic Sobolev inequality (5.1) satisfies*

$$K_0\,(D_0(q) + D_1(q)) \le c \le K_1\,(D_0(q) + D_1(q)).$$

Proof. Without loss of generality, let $m = 0$. The inequality (4.2) may be divided into the two inequalities:

$$\| \, |f_0|^q \, \|_\Psi \le d_0 \int_{-\infty}^0 |f_0'(x)|^q \, d\mu(x), \tag{5.0.5}$$

$$\| \, |f_1|^q \, \|_\Psi \le d_1 \int_0^{+\infty} |f_1'(x)|^q \, d\mu(x), \tag{5.0.6}$$

where f_0 and f_1 are arbitrary smooth functions defined on $(-\infty, 0]$ and $[0, +\infty)$, respectively, with $f_0(0) = f_1(0) = 0$. More precisely, the optimal constants in (5.5)-(5.6) are connected with the optimal constant d in (4.2) by

$$d = \max\{d_0, d_1\}.$$

Now, according to (5.4) with $\nu = \lambda$ being the restriction of μ to $[0, +\infty)$, we have, for the optimal constant d_1,

$$\frac{1}{2} D_1 \le d_1 \le 8\, D_1. \tag{5.0.7}$$

Here we also used an obvious fact that the supremum in the definition of D_1 will not change if one replaces the expression $\mu([x, +\infty))$ by $\mu((x, +\infty))$. Similarly, for the optimal constant d_0, we have

$$\frac{1}{2} D_0 \le d_0 \le 8\, D_0. \tag{5.0.8}$$

By Proposition 4.1, (5.1) implies both (5.5) and (5.6) with $d_i = 42\,c$. Hence, by (5.7)–(5.8), we get

$$\frac{1}{2} \max(D_0, D_1) \le 42\, c.$$

Therefore, $D_0 + D_1 \le 168\, c$ which implies Theorem 5.3 with $K_0 = 1/168$. On the other hand, again by Proposition 4.2, $c \le 144\, d = 144 \max(d_0, d_1)$, so, $c \le 144 \cdot 8 \max(D_0, D_1) \le 1152(D_0 + D_1)$. Thus, one may choose $K_1 = 1152$ which proves Theorem 5.3.

A natural question arising in connection with Theorem 5.3 is how to determine whether or not $D_0(q) + D_1(q)$ is finite in terms of simple conditions on the density p of the measure μ. In the case $q = 2$ this question was studied in [G-R], ([A-B-C-F-G-M-R-S]), and here we follow similar arguments to settle the general case.

Assume the density has the form $p(x) = e^{-U(x)}$ for a twice continuously differentiable function $U : \mathbf{R} \to \mathbf{R}$ (for short, $U \in C^2(\mathbf{R})$). We need the following elementary lemma appearing in [A-B-C-F-G-M-R-S], see pp. 107–109, as a corollary of a more general Proposition 6.4.1 (with further reference to [V-P]).

Lemma 5.4. *Given a function $U \in C^2(\mathbf{R})$ such that $U'(x) > 0$ in $x \geq M$ for some real M, and $\frac{U''(x)}{(U'(x))^2} \to 0$, as $x \to +\infty$, we have, for all $m \in \mathbf{R}$,*

$$\lim_{x\to+\infty} \left(U'(x)\, e^{-U(x)} \int_m^x e^{U(t)}\, dt \right) = 1, \qquad \lim_{x\to+\infty} \left(U'(x)\, e^{U(x)} \int_m^x e^{-U(t)}\, dt \right) = 1.$$

Applying this lemma to $p(x) = e^{-U(x)}$, we obtain that, as $x \to +\infty$:

1) the function $1 - F(x) = \int_m^x e^{-U(t)}\, dt$ is equivalent to $\frac{e^{-U(x)}}{U'(x)}$;

2) the function $\log \frac{1}{1-F(x)}$ is equivalent to $U(x) + \log U'(x)$;

3) $\left(\int_m^x p(t)^{-1/(q-1)}\, dt\right)^{q-1} = \left(\int_m^x e^{U(t)/(q-1)}\, dt\right)^{q-1}$ is equivalent to $\frac{(q-1)^{q-1}\, e^{U(x)}}{(U'(x))^{q-1}}$.

Collecting these asymptotics together, we obtain that the function

$$(1 - F(x)) \log \frac{1}{1 - F(x)} \left(\int_m^x \frac{dt}{p(t)^{1/(q-1)}} \right)^{q-1},$$

appearing in the definition of $D_1(q)$, is equivalent to $(q-1)^{q-1} \frac{U(x) + \log U'(x)}{U'(x)^q}$. We can now summarize (also taking into account an analogous description of the asymptotics near the point $-\infty$):

Theorem 5.5. *Let μ be an absolutely continuous probability measure on $\mathbf{R}$ with density $p(x) = e^{-U(x)}$ where $U \in C^2(\mathbf{R})$. Assume that*

a) $U'(x) > 0$ if $x \geq M$ and $U'(x) < 0$ in $x \leq -M$, for some real M;

b) $\frac{U''(x)}{U'(x)^2} \to 0$, as $|x| \to \infty$.

Then, given $q \in (1, 2]$, the measure μ satisfies the logarithmic Sobolev inequality (5.1) with some finite c if and only if

$$\limsup_{|x|\to\infty} \frac{U(x) + \log |U'(x)|}{|U'(x)|^q} < +\infty. \tag{5.0.9}$$

As an immediate consequence, we get:

Corollary 5.6. *Given $q \in (1, 2]$, the probability measure μ on $\mathbf{R}$ with density $p(x) = const\; e^{-|x|^p}$ satisfies (5.1) if and only if $p \geq \frac{q}{q-1}$.*

The measure μ, with $U(x) = |x|^p + C$, is log-concave, and moreover it has convexity power p (if $p \geq 2$). So, Corollary 5.6 can be proved by a different method, for example, on the basis of Brunn-Minkowski's inequality (cf. [B-L]).

However, Theorem 5.5 allows one to produce a variety of measures which satisfy the logarithmic Sobolev inequality (5.1) but which are not log-concave. Here is one possible example (in a spirit of [A-B-C-F-G-M-R-S]).

Example 5.7. Let U be an even function of class $C^2(\mathbf{R})$ of the form

$$U(x) = |x|^p + \alpha(x) \cos x + C$$

such that $\alpha(x) = |x|^{p-1-\delta}$ for $|x| > 1$, where $p > 2$ and $\delta \in (0,1)$ are fixed parameters. For $x > 1$ we have:

$$U'(x) = px^{p-1} + (p-1-\delta)x^{p-2-\delta} \cos x - x^{p-1-\delta} \sin x,$$

hence $U'(x)$ is equivalent to px^{p-1}, as $x \to +\infty$ (the first term dominates). Next,

$$\begin{aligned} U''(x) &= p(p-1)\, x^{p-2} + (p-1-\delta)(p-2-\delta)\, x^{p-3-\delta} \cos x \\ &\quad - 2(p-1-\delta)\, x^{p-2-\delta} \sin x - x^{p-1-\delta} \cos x. \end{aligned}$$

Now, the last term is of most importance, or, more precisely, we have

$$\frac{U''(x)}{x^{p-1-\delta}} = -\cos x + O(x^{-(1-\delta)}), \quad \text{as } x \to +\infty.$$

Therefore, $U''(x)$ oscillates and attains any values between $-\infty$ to $+\infty$ as the variable x varies along the positive half-axis. Thus, roughly speaking, the corresponding measure μ is very far from being log-concave. On the other hand,

$$\frac{U''(x)}{U'(x)^2} = O(x^{-(p-1+\delta)}),$$

and the property (5.9) is true whenever $p \geq \frac{q}{q-1}$. Hence, for such p the measure μ satisfies the logarithmic Sobolev inequality (5.1).

Remark 5.8. On the basis of Theorem 5.1 and Lemma 5.4 one can easily study the Poincaré-type inequality on the real line,

$$\int_{-\infty}^{+\infty} \left| f - \int_{-\infty}^{+\infty} f \, d\mu \right|^q d\mu \leq c_q \int_{-\infty}^{+\infty} |f'(x)|^q \, d\mu(x), \tag{5.0.10}$$

where $q > 1$ is a parameter, and where f is an arbitrary μ-integrable absolutely continuous function on $\mathbf{R}$ (now, there is no need to restrict ourselves to the bound $q \leq 2$). Under the assumptions a) and b) of Theorem 5.5, we arrive at the following conclusion: the measure μ satisfies (5.10) with some finite c_q if and only if

$$\limsup_{|x| \to \infty} \frac{1}{|U'(x)|^q} < +\infty.$$

Although it resembles (5.9), this property does not depend on q (!) Therefore, all Poincaré-type inequalities (5.10) are equivalent to each other up to constants c_q (of course, under thc above regularity assumptions on the density p).

CHAPTER 6

Exponential integrability and perturbation of measures

Let us return to the abstract situation and to Theorem 1.3 considered in the introductory section. Thus, assume we are given the triple (Ω, μ, Γ) satisfying LS_q-inequality

$$\mathrm{Ent}\,(|f|^q) \leq c \int \Gamma(f)^q \, d\mu \tag{6.0.1}$$

for every function $f \in \mathcal{A}$ (i.e., from the domain of Γ). As before, here q is a given parameter, $1 < q \leq 2$. We are going to generalize Theorem 1.3 by removing the boundedness assumption on the modulus of gradient.

Theorem 6.1. *Under* (6.1), *for every bounded function* $g \in A$ *with* μ*-mean zero,*

$$\int e^g \, d\mu \leq \int e^{cpq^{-q}\,\Gamma(g)^q} \, d\mu, \tag{6.0.2}$$

where $p = \frac{q}{q-1}$ *is conjugate to* q.

As in Theorem 1.3, if we are dealing with probability metric spaces satisfying (6.1), the estimate (6.2) easily extends to non-bounded integrable, locally Lipschitz g. Thus, if $\Gamma(g)^q$ has finite exponential moments, then so does g. Of course, this property cannot be reached on the basis of Theorem 1.3 or the bound (1.3) on the Laplace transform,

$$\int e^{\lambda g} \, d\mu \leq \exp\left\{\frac{c}{q^q(q-1)}\,\lambda^q\right\}, \quad \lambda > 0. \tag{6.0.3}$$

Note that, if $\Gamma(g) \leq 1$ as in (6.3), the estimate (6.2) yields somewhat different:

$$\int e^{\lambda g} \, d\mu \leq \exp\left\{\frac{cp}{q^q}\,\lambda^q\right\}, \quad \lambda > 0.$$

Here the constant in the exponent $\frac{p}{q^q} = \frac{q}{q^q(q-1)}$ is $q \leq 2$ times worse than that in (6.3). Hence, Theorem 1.3 still has a small advantage over Theorem 6.1 in case $\Gamma(g) \leq 1$.

Proof of Theorem 6.1. The argument is similar to the one given in [B-G1] for the particular case $q = 2$. Given a number $\alpha > 0$, define $\beta > 0$ by

$$\int e^{\alpha\,\Gamma(g)^q} \, d\mu = e^{\beta}. \tag{6.0.4}$$

We use the following well-known representation for the entropy functional: for every measurable, bounded function $w \geq 0$ on Ω,

$$\text{Ent}\,(w) \; = \sup_{\int e^h \, d\mu \leq 1} \int wh \, d\mu. \tag{6.0.5}$$

Here the supremum may be taken over all measurable bounded function h on Ω such that $\int e^h \, d\mu \leq 1$.

Let $w = e^g$ and let $h = \alpha\Gamma(g)^q - \beta$ so that, by (6.4), $\int e^h \, d\mu = 1$. Hence, by (6.5), we get

$$\int e^g \left(\alpha\Gamma(g)^q - \beta\right) d\mu \leq \text{Ent}\,(e^g).$$

On the other hand, by the log-Sobolev inequality (6.1) with $f = e^{g/q}$,

$$\text{Ent}\,(e^g) \leq \frac{c}{q^q} \int \Gamma(g)^q \, e^g \, d\mu. \tag{6.0.6}$$

Combining the both estimates gives

$$\left(\alpha - \frac{c}{q^q}\right) \int \Gamma(g)^q \, e^g \, d\mu \leq \beta \int e^g \, d\mu.$$

Assuming $\alpha > \frac{c}{q^q}$, the left hand side can further be estimated from below by applying once more (6.6). Then we get

$$\text{Ent}\,(e^g) \leq C\beta \int e^g \, d\mu, \quad C = \frac{1}{\frac{\alpha q^q}{c} - 1}.$$

Now, given $\lambda \in [0, 1]$, let us write this inequality for the function λg in the place of g. Then,

$$\text{Ent}\,(e^{\lambda g}) \leq C\beta(\lambda) \int e^{\lambda g} \, d\mu, \quad C = \frac{1}{\frac{\alpha q^q}{c} - 1}, \tag{6.0.7}$$

where $\beta(\lambda)$ should be defined according to (6.4) as

$$\int e^{\alpha\lambda^q \, \Gamma(g)^q} \, d\mu = e^{\beta(\lambda)}.$$

Since $\lambda \in [0, 1]$, by Hölder's inequality, $\int e^{\alpha\lambda^q \, \Gamma(g)^q} \, d\mu \leq \left(\int e^{\alpha \, \Gamma(g)^q} \, d\mu\right)^{\lambda^q}$, so $\beta(\lambda) \leq \beta\lambda^q$, where recall that $\beta(1) = \beta$. Therefore, from (6.7),

$$\text{Ent}\,(e^{\lambda g}) \leq C\beta \, \lambda^q \int e^{\lambda g} \, d\mu, \quad C = \frac{1}{\frac{\alpha q^q}{c} - 1}. \tag{6.0.8}$$

Now we can proceed as in the proof of Theorem 1.3. Write the Laplace transform on g as $\int e^{\lambda g} \, d\mu = e^{\lambda v(\lambda)}$. The function v is smooth in $\lambda > 0$ and satisfies $v(0+) = 0$ in view of the assumption that g has μ-mean zero. Recall that $\text{Ent}\,(e^{\lambda g}) = \lambda^2 v'(\lambda) \int e^{\lambda g} \, d\mu$. Hence, by (6.8), $\lambda^2 v'(\lambda) \leq C\beta \, \lambda^q$ and

$$v(\lambda) = \int_0^\lambda v'(t) \, dt \leq C\beta \int_0^\lambda t^{q-2} \, dt = \frac{C\beta}{q-1} \, \lambda^{q-1}.$$

Thus, we have obtained the estimate $\int e^{\lambda g} \, d\mu \leq \exp\{\frac{C\beta}{q-1} \, \lambda^q\}$ which for $\lambda = 1$ becomes (recalling the definition of β and C)

$$\int e^g \, d\mu \leq e^{\frac{C\beta}{q-1}} = \left(\int e^{\alpha \, \Gamma(g)^q} \, d\mu\right)^{\frac{C}{q-1}} = \left(\int e^{\alpha \, \Gamma(g)^q} \, d\mu\right)^{\frac{1}{(q-1)(\frac{\alpha q^q}{c} - 1)}}. \tag{6.0.9}$$

It remains to equalize

$$(q-1)\left(\frac{\alpha q^q}{c}-1\right)=1$$

from which we easily find that $\alpha = \frac{cp}{q^q}$. For this value (6.9) coincides with (6.2), and this proves Theorem 6.1.

Next we briefly investigate which perturbation of measures preserve LS_q-inequality. This will be useful for further purposes.

First of all we mention that following the idea [**H-S**] in the general setting one has

Proposition 6.2. *Suppose $\mu \in \mathrm{LS}_q(c)$ and let $d\mu_\rho \equiv \rho d\mu$ for some strictly positive bounded density ρ. Then $\mu_\rho \in \mathrm{LS}_q(c')$ with $c' \equiv c\exp\{\sup(\log\rho) - \inf(\log\rho)\}$.*

For the proof it is sufficient to notice that the nonlinear expression in μ_ρ can be achieved as an infimum of expectation of nonnegative function as follows

$$\mu_\rho f^q \log\frac{f^q}{\mu f^q} = \inf_{t>0}\mu_\rho\left(f^q\log f^q - f^q\log t - f^q + t\right)$$

Now we can remove the density ρ and use $\mu \in \mathrm{LS}_q(c)$ and finally replace the expectation $\mu|\nabla f|_q^q$ by the one we need. The total price for these operations equals to the exponential of $\sup\log\rho - \inf\log\rho$. (We mention that in certain situation when more structure is available Proposition 6.2 can be strengthened, see e.g. [**G-Z**]. Naturally similar arguments works directly for the related Poincaré inequality SG_q.)

The second perturbation formula applies to the situation in which we have a local gradient and involves unbounded and possibly locally singular perturbations to which Theorem 5.5 does not apply. This includes for example perturbations where

$$\log\rho = \sum_j a_j V(\omega - b_j) + const$$

with a singular function V, (singularity may be possibly growing with the dimension of the space), and certain infinite sequences of points a_j and b_j.

Proposition 6.3. *Suppose $\mu \in \mathrm{LS}_q(c)$ and let $d\mu_\rho \equiv \rho d\mu$ for a positive density ρ. If*

$$\mu_\rho\left(\exp\left\{\frac{2^{q-1}q^{-q}c}{\varepsilon}|\nabla\log\rho|_q^q\right\}\right) < \infty$$

for some constant $\varepsilon \in (0,1)$ and μ_ρ satisfies SG_q with a constant $m' \in (0,\infty)$, then $\mu_\rho \in \mathrm{LS}_q(c')$ for some $c' \in (0,\infty)$.

Proof. Inserting $\rho^{1/q}f$, defined with bounded smooth cylinder function f normalized so that $\mu_\rho|f|^q = 1$, into $\mathrm{LS}_q(c)$ for the measure μ after simple arguments (utilizing the Leibnitz rule for the local gradient) one arrives at the following bound

$$\mu_\rho|f|^q\log|f|^q \le 2^{q-1}c\mu_\rho|\nabla f|_q^q + 2^{q-1}q^{-q}c\mu_\rho(|f|^q|\nabla\log\rho|_q^q) \tag{6.0.10}$$

Next utilizing Young inequality $ab \leq a \log a + 2e^b$, (as e.g. in [**L6**]), we bound the second term on the right hand side of (6.0.10) as follows

$$2^{q-1}q^{-q}c\mu_\rho(|f|^q|\nabla \log \rho|_q^q) \leq \varepsilon\mu_\rho|f|^q \log|f|^q + 2\mu_\rho\left(\exp\left\{\frac{2^{q-1}q^{-q}c}{\varepsilon}|\nabla \log \rho|_q^q\right\}\right)$$

with arbitrary $\varepsilon \in (0,1)$. Inserting this into (6.0.10), after simple transformations we arrive at

$$\mu_\rho|f|^q \log|f|^q \leq \frac{2^{q-1}c}{1-\varepsilon}\mu_\rho|\nabla f|_q^q + \frac{2}{1-\varepsilon}\mu_\rho\left(\exp\left\{\frac{2^{q-1}q^{-q}c}{\varepsilon}|\nabla \log \rho|_q^q\right\}\right) \tag{6.0.11}$$

To complete the proof it is sufficient to use the adaptation of Rothaus lemma proven in Section 3 together with the assumed SG_q for the perturbed measure. Finally, after this stage we can relax our initial assumption $\mu_\rho|f|^q = 1$.

CHAPTER 7

LS_q-inequalities for Gibbs measures with Super Gaussian Tails.

In this section we show that LS_q is satisfied for a large class of nontrivial infinite dimensional measure. Our arguments generalize the ones developed to show LS_2 and to keep the size of this section reasonable we concentrate only on the aspects which require modification. (We refer to [**G-Z**] for a comprehensive description of LS_2 case.)

For the present purposes it is convenient to define the infinite dimensional measures by introducing an a'priori given family of its regular conditional expectations which are given in terms of compatible family of probability kernels involving essentially finite dimensional integration as follows. For $\omega \in \mathbf{R}^{\mathbb{Z}^d}$ and a finite set $\Lambda \subset\subset \mathbb{Z}^d$ we define

$$E_\Lambda^\omega(f) \equiv \delta_\omega\left(\frac{\int f\ e^{-U_\Lambda} d\nu_{r,\Lambda}}{\int e^{-U_\Lambda} d\nu_{r,\Lambda}}\right) \tag{7.0.1}$$

where $\nu_{r,\Lambda} \equiv \otimes_{i\in\Lambda}\nu_r(dx_i)$ and

$$U_\Lambda \equiv \sum_{X\cap\Lambda\neq\emptyset} \Phi_X \tag{7.0.2}$$

where $\Phi_X(\omega) \equiv \psi_X(\omega_X)$ for a twice continuously differentiable function ψ_X on $\mathbf{R}^X$ and $\omega_X \equiv (\omega_i)_{i\in X}$; δ_ω denotes the point mass concentrated at $\omega \in \mathbf{R}^{\mathbb{Z}^d}$. Such family called a local specification can be defined for any measure $\nu_r(dy) \equiv \alpha_r^{-1}e^{-|y|^r}dy$, $r \in (1,\infty)$, on $\mathbf{R}$ and the interaction $\Phi \equiv \{\Phi_X\}_{X\subset\subset\mathbb{Z}^d}$ such that for any $\Lambda \subset\subset \mathbb{Z}^d$ one has

$$0 < \int e^{-U_\Lambda} d\nu_{r,\Lambda} < \infty \tag{7.0.3}$$

For simplicity we assume that

$$||\Phi|| \equiv \sup_{i\in\mathbf{Z}^d} \sum_{X\subset\subset\mathbf{Z}^d X\ni i,|X|>1} |X|\cdot||\Phi_X||_\infty < \infty$$

although as one can see our results hold true in greater generality. By the very definition of the local specification $E_\Lambda^\omega(f)$ is measurable with respect the smallest σ-algebra generated by the coordinate functions $\{\omega_i : i \in \mathbb{Z}^d \setminus \Lambda\}$ and one has the following compatibility condition satisfied for any $\widetilde{\Lambda} \subset \Lambda$

$$E_\Lambda^\omega\left(E_{\widetilde{\Lambda}}^{\cdot}(f)\right) = E_\Lambda^\omega(f)$$

for any bounded measurable function f.

In this section we restrict ourselves to the case of super - Gaussian tails, that is we assume that $r \in (2, \infty)$. Under this condition simple perturbation arguments (mentioned in the previous section) show that E_Λ inherits the LS_q property with some finite constant, that is in shorthand notation

$$E_\Lambda \in \mathrm{LS}_q(C_\Lambda)$$

with some finite constant $C_\Lambda \in (0, \infty)$.

Later on we assume that local specification satisfies a mixing condition as the one assumed in the proof of the LS_2 and which can be formulated (in a strong form) as follows

Strong Mixing Condition

$$|E_\Lambda\left((f - E_\Lambda f)(g - E_\Lambda g)\right)| \leq A(\nabla f, \nabla g) e^{-M\ dist(\Lambda_f, \Lambda_g)}$$

where $M \in (0, \infty)$ and $A(\nabla f, \nabla g)$ are constants independent of a finite set $\Lambda \subset\subset \mathbb{Z}^d$ and $dist(\Lambda_f, \Lambda_g)$ is the Euclidean distance between smallest sets Λ_f and Λ_g such that f and g are measurable with respect to smallest σ-algebra generated by coordinate functions $\{\omega_i : i \in \Lambda_f\}$ and $\{\omega_i : i \in \Lambda_g\}$, respectively; (see e.g. [**G-Z**]).

We mention that for example such condition is satisfied under the Dobrushin uniqueness ([**G-Z**]). Let μ be a Gibbs measure for the local specification $\{E_\Lambda\}_{\Lambda \subset\subset \mathbb{Z}^d}$ that is we have

$$\mu E_\Lambda f \ = \mu f$$

for any bounded measurable function f ; here and later on we use the notation $\mu f \equiv \int f d\mu$.

We will prove that under suitable conditions on the local specification, the Gibbs measure satisfies LS_q . To show this, we note first that under our general assumption this inequality is true for any kernel E_Λ , (one apply here similar perturbation lemma as used in case of LS_2, see e.g. [**G-Z**]). The idea of the proof is to use a suitable telescopic representation of relative entropy associated to an appropriate sequence of transition matrices (or conditional expectations) as follows. Given a sequence of transition matrices, E_n, $n \in \mathbb{N}$ such that

$$\mu E_n F = \mu F \tag{7.0.4}$$

for a probability measure μ, we set $f_n \equiv (E_n f_{n-1}^q)^{\frac{1}{q}}$, with $f_0 \equiv f$ for a nonnegative function f. Assuming that

$$\lim_{n \to \infty} f_n = \mu f \tag{7.0.5}$$

we note that

$$\mu f^q \log \frac{f^q}{\mu f^q} = \sum_{n \in \mathbb{N}} \mu \left(E_n f_{n-1}^q \log \frac{f_{n-1}^q}{E_n f_{n-1}^q} \right) \tag{7.0.6}$$

Hence, if with some constant $C \in (0, \infty)$ independent of $n \in \mathbb{N}$ one has

$$E_n f_{n-1}^q \log \frac{f_{n-1}^q}{E_n f_{n-1}^q} \leq C\ E_n |\nabla_n f_{n-1}|^q \tag{7.0.7}$$

(with a gradient ∇_n associated to the integration variables of E_n), using (7.0.6) and (7.0.4), we obtain

$$\mu f^q \log \frac{f^q}{\mu f^q} \leq C \sum_{n\in\mathbb{N}} \mu \, |\nabla_n f_{n-1}|^q \tag{7.0.8}$$

Since $f_{n-1} \equiv (E_{n-1} f_{n-2}^q)^{\frac{1}{q}}$, if the probability measures E_{n-1} do not depend on the n-th variables (as in the product case), by Minkowski inequality one gets

$$|\nabla_n f_{n-1}|^q \leq E_{n-1} |\nabla_n f_{n-2}|^q \tag{7.0.9}$$

and by induction, (if E_k , $k \leq n-1$, are independent of the differentiation variable in question), one arrives at

$$|\nabla_n f_{n-1}|^q \leq E_{n-1} |\nabla_n f_{n-2}|^q \leq E_{n-1} ... E_1 |\nabla_n f|^q \tag{7.0.10}$$

Thus combining (7.0.8) - (7.0.10), we arrive at

Theorem 7.1. (Product property of LS_q) If $\mu \equiv \otimes_{n\in\mathbb{N}} \mu_n$ with $\mu \in \mathrm{LS}_q(C)$, then

$$\mu f^q \log \frac{f^q}{\mu f^q} \leq C \sum_{n\in\mathbb{N}} \mu \, |\nabla_n f|^q \tag{7.0.11}$$

If E_{n-1} depends on the n-th variable, (as we would need to have to be able to include the Gibbs and PCA measures), we use similar arguments as described in [**G-Z**]. One introduces E_{n-1} defined as an appropriate product and/or convolution of probability kernels $E_{\Delta+j}$ associated to suitable translations of a finite cube Δ. To estimate $|\nabla_n f_{n-1}|^q$ it is necessary to get an estimate on $|\nabla_i (E_\Lambda F^q)^{\frac{1}{q}}|$ defined for a finite set $\Lambda \subset \mathbb{Z}^d$. One notes that for $i \notin \Lambda$, we have

$$|\nabla_i (E_\Lambda F^q)^{\frac{1}{q}}| = |\frac{1}{q} (E_\Lambda F^q)^{\frac{1}{q}-1} \nabla_i (E_\Lambda F^q)| \tag{7.0.12}$$

Using the definition of E_Λ as a perturbation of a product measure we can compute the derivative on the right hand side as follows

$$|\nabla_i (E_\Lambda F^q)| = |E_\Lambda (F^q (-\nabla_i U_\Lambda + E_\Lambda \nabla_i U_\Lambda)) + q E_\Lambda (F^{q-1} \nabla_i F)| \tag{7.0.13}$$

Since the covariance vanishes if one of the involved functions is a constant, setting $F_s \equiv sF + (1-s) E_\Lambda F$, we get

$$|E_\Lambda (F^q (-\nabla_i U_\Lambda + E_\Lambda \nabla_i U_\Lambda) | = | \int_0^1 ds \, \frac{d}{ds} E_\Lambda \left(F_s^q (-\nabla_i U_\Lambda + E_\Lambda \nabla_i U_\Lambda) \right) |$$

$$\begin{aligned} &\leq \; q \left(E_\Lambda F_s^q | - \nabla_i U_\Lambda + E_\Lambda \nabla_i U_\Lambda |^p \right)^{\frac{1}{p}} \left(E_\Lambda |F - E_\Lambda F|^q \right)^{\frac{1}{q}} \\ &\leq \; q \; var_\Lambda (\nabla_i U_\Lambda) \; \sup_{s\in[0,1]} \left(E_\Lambda F_s^q \right)^{\frac{1}{p}} \left(E_\Lambda |F - E_\Lambda F|^q \right)^{\frac{1}{q}} \end{aligned}$$

Now using the $\mathbb{L}_q$ spectral gap inequality for the measure E_Λ, (which follows from LS_q by general arguments of Proposition 2.1), we arrive at

(7.0.14)

$|E_\Lambda (F^q (-\nabla_i U_\Lambda + E_\Lambda \nabla_i U_\Lambda) | \leq$

$$\leq q\,(E_\Lambda F^q)^{1-\frac{1}{q}} \left[2\left(\frac{4C_\Lambda}{\log 2}\right)^{\frac{1}{q}} var_\Lambda(\nabla_i U_\Lambda) \sum_{j\in\Lambda} (E_\Lambda\,|\nabla_j F|^q)^{\frac{1}{q}} \right]$$

Combining (7.0.12) - (7.0.14), via simple arguments (similar to the ones used in [**G-Z**] for the case of LS_2), one arrives at a bound of the following form

$$|\nabla_i(E_\Lambda F^q)^{\frac{1}{q}}| \leq |(E_\Lambda|\nabla_i F|^q)^{\frac{1}{q}} + C'\ var_\Lambda(\nabla_i U_\Lambda) \sum_{j\in\Lambda} (E_\Lambda|\nabla_j F|^q)^{\frac{1}{q}} \tag{7.0.15}$$

with some positive constant C' and $var_\Lambda(\nabla_i U_\Lambda)$ denoting the total variation of $\nabla_i U_\Lambda$ with respect to the coordinates indexed by points in the finite set Λ. Note that in case $i \in \Lambda$, the derivative on the right hand side vanishes by definition of our kernel E_Λ. If $var_\Lambda(\nabla_i U_\Lambda)$ is sufficiently small, the inequality (7.0.15) is sufficient to complete the proof of LS_q for the Gibbs measure μ. However if the local specification satisfies the strong mixing property, this smallness requirement can be essentially weakened.

To take the full advantage of mixing, one may follow the same inductive idea of conditioning as used in the case of LS_2 inequalities (see e.g. [**G-Z**]) to show appropriate relations for the present case. That is given a finite set Λ we choose a subset $\widetilde{\Lambda} \subset \Lambda$ and then using the compatibility property of kernels we note first that

$$|\nabla_i(E_\Lambda F^q)^{\frac{1}{q}}| = |\nabla_i(E_\Lambda E_{\widetilde{\Lambda}} F^q)^{\frac{1}{q}}| \equiv |\nabla_i(E_\Lambda \widetilde{F}^q)^{\frac{1}{q}}|$$

Next we apply the above arguments with the function $\widetilde{F}$ where by a proper choice of the set $\widetilde{\Lambda}$, (so that $\widetilde{F}$ is localised possibly far from $\nabla_i U_\Lambda$), we can take the advantage of the mixing condition. Finite iteration of this procedure result with the improved sweeping out relations of the form

(Sweeping out relations)

$$|\nabla_i(E_\Lambda F^q)^{\frac{1}{q}}| \leq |(E_\Lambda|\nabla_i F|^q)^{\frac{1}{q}} + \sum_{j\in\Lambda} C_{ij}(E_\Lambda|\nabla_j F|^q)^{\frac{1}{q}} \tag{7.0.16}$$

where $C_{ij} \in [0,\infty)$ decay in a suitable way with the growth of $dist(i,j)$ (see [**G-Z**] for details).

Using the sweeping out relations one shows the following contraction bound involving the transitions matrices

$$|\nabla(E_{n-1}g^q)^{\frac{1}{q}}|_q^q \leq \lambda E_{n-1}|\nabla g|_q^q \tag{7.0.17}$$

with a constant $\lambda \in (0,1)$ independent of $n \in \mathbf{N}$, where $|\nabla g|_q^q \equiv \sum_j |\nabla_j g|^q$. With $g = f_{n-2}$ and the help of simple induction, this yields

$$|\nabla(E_{n-1}f_{n-2})^{\frac{1}{q}}|_q^q \leq \lambda^{n-1} E_{n-1}|\nabla f|_q^q \tag{7.0.18}$$

Application of (7.0.18) together with (7.0.11) implies the following LS_q inequality

$$\mu f^q \log \frac{f^q}{\mu f^q} \leq \frac{C_q}{1-\lambda}\mu|\nabla f|_q^q$$

This completes the proof of the following result.

Theorem 7.2. (LS_q for Gibbs measures) *Let μ be a Gibbs measure corresponding to the local specification $\{E_\Lambda\}_{\Lambda\subset\subset\mathbb{Z}^d}$ defined by (7.0.1)-(7.0.3) with the reference product measure with super-Gaussian tails. If the local specification satisfies the mixing condition, then there is a constant $c_q \in (0,\infty)$ such that*

$$(\mathrm{LS}_q(\mu)) \qquad \mathrm{Ent}_\mu(f) \le c_q \cdot \mu|\nabla f^{\frac{1}{q}}|_q^q$$

for all nonnegative functions f for which the right hand side is finite.

Remark 7.3. We also remark that essentially the same arguments work also in the case of PCA measures. The case of perturbations by unbounded potentials can also be included, though they require lengthy and technically slightly more involved description.

The above theorem generalises a result of [**B-L**] where a corresponding result was proven under a very strong log-concavity assumptions of the finite dimensional measures. That is, for a probability measure having a density $\exp(-V)/Z$ with respect to a finite dimensional Lebesgue measure with

$$V(x)+V(y)-2V(\frac{x+y}{2}) \ge const\ ||x-y||^p$$

for any x and y, such probability measure satisfies LS_q (with dual norm on the tangent space).

We remark that in the considered case of local specification with super Gaussian tails the following Generalised Nash inequality was proved in [**Z2**] for the corresponding Gibbs measures following natural adaptation of classical Sobolev inequality for probability measures by [**Ros**]

$$(\mathrm{GN}_q) \qquad \mu f^2 \log\frac{f^2}{\mu f^2} \le C\frac{2^q}{q^q}\sum_j \left(\mu|\nabla_j f|^2\right)^{\frac{q}{2}} \left(\mu(f-\mu f)^2\right)^{\frac{2-q}{2}}$$

It was proved there that such inequality produces the correct exponential bounds for Lipschitz function.

The following result explains the relation of GN_q inequality and LS_q

Theorem 7.4. *Suppose $r \in [2,\infty)$ and $\frac{1}{r}+\frac{1}{q}=1$. If LS_q is satisfied with a coefficient $C \in (0,\infty)$ and the following spectral gap inequality holds*

$$(\mathrm{SG}_2) \qquad \mu(f-\mu f)^2 \le m^{-1}\mu|\nabla f|_2^2$$

with some constant $m \in (0,\infty)$, then the Generalised Nash inequality (GN_q)is true with some constant $C \in (0,\infty)$ for all nonnegative functions f for which the right hand side is finite.

Remark 7.5. Note that, although SG_2 follows from SG_q in finite dimension (and the last is an abstract consequence of LS_q), in infinite dimensional setting we need to be more careful as l_q and l_2 norms are not equivalent.

We also remark that if dimension of the underlying space is finite, then GN_q, $q \in (1,2)$, implies LS_2 . But in infinite dimensional situation this is in general not true.

Proof of Theorem 7.4. Suppose LS_q is satisfied with $q \in (1,2)$. Then substituting $f^{\frac{2}{q}}$ one gets

$$\mu f^2 \log \frac{f^2}{\mu f^2} \leq C\mu|\nabla f^{\frac{2}{q}}|_q^q = C\mu \sum_j |\nabla_j f^{\frac{2}{q}}|^q \tag{7.0.19}$$

We have

$$\mu \sum_j |\nabla_j f^{\frac{2}{q}}|^q = \frac{2^q}{q^q}\mu\left(f^{2-q}\left(\sum_j |\nabla_j f|^q\right)\right)$$

Since $q \leq 2$ and using Hölder inequality, we get

$$\mu \sum_j |\nabla_j f^{\frac{2}{q}}|^q \leq \frac{2^q}{q^q}\left(\mu|\nabla f|_q^2\right)^{\frac{q}{2}}\left(\mu f^2\right)^{\frac{2-q}{2}} \leq \frac{2^q}{q^q}\sum_j \left(\mu|\nabla_j f|^2\right)^{\frac{q}{2}}\left(\mu f^2\right)^{\frac{2-q}{2}} \tag{7.0.20}$$

Next note that

$$\mu f^2 \log \frac{f^2}{\mu f^2} \leq \mu(f-\mu f)^2 \log \frac{(f-\mu f)^2}{\mu(f-\mu f)^2} + 2\mu(f-\mu f)^2 \tag{7.0.21}$$

and, if the Poincaré inequality is true, we have

$$\begin{aligned} \mu(f-\mu f)^2 &= \left(\mu(f-\mu f)^2\right)^{\frac{q}{2}}\left(\mu(f-\mu f)^2\right)^{\frac{2-q}{2}} \\ &\leq \left(m^{-1}\mu|\nabla f|_2^2\right)^{\frac{q}{2}}\left(\mu(f-\mu f)^2\right)^{\frac{2-q}{2}} \end{aligned} \tag{7.0.22}$$

Thus applying (7.0.19) - (7.0.20) with $|f-\mu f|$ instead of f to estimate the first term on the right hand side of (7.0.21) and using (7.0.22) to estimate the second one, we arrive at the following bound

$$\mu f^2 \log \frac{f^2}{\mu f^2} \leq \left(\frac{2^q}{q^q}C + 2m^{-\frac{q}{2}}\right)\sum_j \left(\mu|\nabla_j f|^2\right)^{\frac{q}{2}}\left(\mu(f-\mu f)^2\right)^{\frac{2-q}{2}}$$

□

CHAPTER 8

LS_q-inequalities and Markov semigroups

In this section we discuss consequences of LS_q for the semigroups preserving the related measure. The first result provides a relation to the ultracontractivity property of a semigroup generated by a Dirichlet operator satisfying

$$\mu(f\mathbf{L}f) = -\mu|\nabla f|_2^2$$

The result holds in a general setting including both discrete as well as continuous spaces.

Theorem 8.1. *Suppose the dimension of the underlying space is finite and assume that $r \in (2,\infty)$ and $\frac{1}{r} + \frac{1}{q} = 1$. If LS_q is satisfied , then the following inequality is true for any $p \in (2,\infty)$ with generator $\mathbf{L}$ corresponding to the L_2 - Dirichlet form*

$$(\mathrm{GLS}_q) \qquad \mu f^p \log \frac{f^p}{\mu f^p} \leq -p\varepsilon(p)\, \mu\left(f^{p-1}\mathbf{L}f\right) + p\eta(p)\, \mu f^p$$

where

$$\varepsilon(p) \equiv \tfrac{a}{(p-1)}p^{-\delta} \quad \text{and} \quad \eta(p) \equiv bp^{-2+\delta q/(2-q)}$$

with some constants $a,b \in (0,\infty)$ and $\delta \in (0, \frac{2-q}{q})$, for all nonnegative functions f for which the right hand side is finite.

Therefore the semigroup $P_t \equiv e^{t\mathbf{L}}$ is ultracontractive, that is for any $p \in [1,\infty)$ and $t > 0$ the operator $P_t : L^p \longrightarrow L^\infty$ is bounded.

Proof. If the dimension of the space is finite, by similar arguments as in the proof of Theorem 7.4 , one gets

$$\mu f^2 \log \frac{f^2}{\mu f^2} \leq C' \left(\mu|\nabla f|_2^2\right)^{\frac{q}{2}} \left(\mu f^2\right)^{\frac{2-q}{2}}$$

and hence for any constant $\alpha \in (0,\infty)$ we have

$$\mu f^2 \log \frac{f^2}{\mu f^2} \leq C'\frac{q}{2}\alpha\, \mu|\nabla f|_2^2 + C'\frac{2-q}{q}\alpha^{-\frac{q}{2-q}}\, \mu f^2$$

Substituting $f^{\frac{p}{2}}$, with $p \in (2,\infty)$, and using the well known inequality

$$\mu|\nabla f^{\frac{p}{2}}|_2^2 \leq -\frac{p^2}{4(p-1)}\mu\left(f^{p-1}\mathbf{L}f\right)$$

for the Dirichlet generator $\mathbf{L}$, we get , with appropriate constants $a,b \in (0,\infty)$ independent of p , the following bound

$$\mu f^p \log \frac{f^p}{\mu f^p} \leq -a\frac{p^2}{4(p-1)}\alpha\, \mu\left(f^{p-1}\mathbf{L}f\right) + b\alpha^{-\frac{q}{2-q}}\, \mu f^p$$

We will choose

$$\alpha \equiv \alpha(p) \equiv p^{-\delta}$$

with some $\delta \in (0, \frac{2-q}{q})$ and with this choice we have the following inequality

$$\mu f^p \log \frac{f^p}{\mu f^p} \leq -p\varepsilon(p)\ \mu\left(f^{p-1}\mathbf{L}f\right) + p\eta(p)\ \mu f^p$$

where

$$\varepsilon(p) \equiv a\frac{p^{1-\delta}}{4(p-1)} \quad \text{and} \quad \eta(p) \equiv bp^{-1+\frac{\delta q}{2-q}}$$

It is not difficult to check that

$$\int_2^\infty p^{-1}\varepsilon(p)dp < \infty \quad \text{and} \quad \int_2^\infty p^{-1}\eta(p)dp < \infty$$

which following [**G1**] implies the ultracontractivity of the semigroup $P_t = e^{t\mathbf{L}}$.

Remark 8.2. We recall following [**D-S**] (see also [**D**] and [**Ros**]) that in finite dimensions the distributions with super-Gaussian tails define Dirichlet generators of ultracontractive semigroups.

8.1. Ergodicity for nonlinear semigroups

Let $T_t \equiv e^{t\mathcal{L}}$ be a semigroup with generator $\mathcal{L}$ defined by

$$\mathcal{L} \equiv \sum_j \mathcal{L}_j$$

where $\mathcal{L}_j$ satisfies

$$\mu\left(g\mathcal{L}_j(f)\right) = -\mu\left(\nabla_j g \cdot \mathbf{v}_j(|\nabla_j f|)\nabla_j f\right)$$

with a nonnegative function $\mathbf{v}_j$.

Theorem 8.3. ($\mathbf{L}_2$ ergodicity) *If the following Spectral Gap inequality is satisfied*

$$\text{(SG)} \qquad m \cdot \mu|g - \mu g|^q \leq \sum_j \mu\left(\mathbf{v}_j(|\nabla_j g|) \cdot |\nabla_j g|^2\right)$$

with some constants $q \in (1, \infty)$ and $m \in (0, \infty)$, then following estimate is true

$$\mu(T_t f - \mu f)^2 \leq D \cdot t^{-\beta} \tag{8.1.1}$$

with some constants $D, \beta \in (0, \infty)$ for all nonnegative functions f in. ∘

Remark 8.4. Recall that by Theorem 2.1, SG_q follows from LS_q, (but if the dimension of the underlying space is infinite SG_q does not imply SG_2).

Proof. Let $f_t \equiv T_t f$, for a smooth bounded function f such that $\mu f = 0$. We have formally

$$\frac{d}{dt}\mu f_t^2 = 2\mu f_t \mathcal{L} f_t = -2\sum_j \mu\left(\mathbf{v}_j(|\nabla_j f_t|) \cdot |\nabla_j f_t|^2\right) \tag{8.1.2}$$

(Since a'priori the question whether f_t is in the domain of $\mathcal{L}$ may be problematic, one may need to use here appropriate approximation arguments.) If the following spectral gap inequality is satisfied

$$m \cdot \mu|g - \mu g|^q \leq \sum_j \mu\left(\mathbf{v}_j(|\nabla_j g|) \cdot |\nabla_j g|^2\right)$$

then one has

$$\frac{d}{dt}\mu f_t^2 \leq -2m \cdot \mu|f_t|^q \tag{8.1.3}$$

Suppose $q \leq 2$, then we use Hölder inequality to get the following bound

$$\mu|f_t|^2 \leq (\mu f_t^q)^{\frac{1}{u}} \left(\mu f_t^{(2-s)w}\right)^{\frac{1}{w}} \leq (\mu f_t^q)^{\frac{1}{u}} \left(\mu f^{(2-s)w}\right)^{\frac{1}{w}}$$

with arbitrary $s \in (0,2)$ and $u, w \in (1, \infty)$ satisfying $su = q$, $\frac{1}{u} + \frac{1}{w} = 1$. Hence we have

$$\left(\frac{\mu|f_t|^2}{\left(\mu f^{(2-s)w}\right)^{\frac{1}{w}}}\right)^u \leq \mu f_t^q \tag{8.1.4}$$

We use this to bound the right hand side of (8.1.3) as follows

$$\frac{d}{dt}\mu f_t^2 \leq -2m \cdot \left(\frac{\mu|f_t|^2}{\left(\mu f^{(2-s)w}\right)^{\frac{1}{w}}}\right)^u \tag{8.1.5}$$

Solving this differential inequality, we arrive at

$$\begin{aligned} \mu f_t^2 &\leq \left(\left(\mu f^2\right)^{1-u} + 2m(u-1) \cdot \left(\frac{1}{\left(\mu f^{(2-s)w}\right)^{\frac{1}{w}}}\right)^u t\right)^{-\frac{1}{u-1}} \\ &\leq \left(\frac{\left(\mu f^{(2-s)w}\right)^{\frac{u}{w}}}{2m(u-1)}\right)^{\frac{1}{u-1}} \cdot t^{-\frac{1}{u-1}} \end{aligned} \tag{8.1.6}$$

whence the bound follows.

Theorem 8.5. (Ergodicity in the Entropy Sense) *Suppose*

$$\mu\left(f\mathcal{L}(f)\right) \equiv -\sum_{j\in\Lambda} \mu\left(\nabla_j f \cdot \mathbf{v}_j(|\nabla_j f|)\nabla_j f\right)$$

with $v_j(y) \geq |y|^{\kappa-2}$ *for some* $\kappa \geq 2$. *If the following* LS_q *inequality is satisfied*

$$\mathrm{Ent}_\mu(f) \leq q^q C \cdot \mu|\nabla f^{\frac{1}{q}}|_q^q \tag{LS_q}$$

with $q \in (1,2)$, $\frac{1}{q} + \frac{1}{\kappa} = 1$, *and some constant* $C \in (0, \infty)$ *and* $|\Lambda| < \infty$, *then following estimate is true*

$$\mathrm{Ent}_\mu(f_t) \leq |\Lambda| C^{-\frac{\kappa}{\kappa-q}} \left(\frac{\kappa-q}{q}\right)^{\frac{q}{\kappa-q}} \cdot t^{-\frac{q}{\kappa-q}} \tag{8.1.7}$$

for all nonnegative functions f.

Proof. Let $f_t \equiv T_t f \equiv e^{t\mathcal{L}}$, for a smooth bounded function $f \geq 0$. Setting $S_t \equiv \mathrm{Ent}_\mu(f_t)$, we have

$$\frac{d}{dt}S_t = \mu\mathcal{L}f_t \log\frac{f_t}{\mu f_t} = -\sum_j \mu\left(f_t^{-1}\mathbf{v}_j(|\nabla_j f_t|)\cdot|\nabla_j f_t|^2\right) \tag{8.1.8}$$

If $\mathbf{v}_j(y) \geq |y|^{\kappa-2}$, then we have

$$\mu\left(f_t^{-1}\mathbf{v}_j(|\nabla_j f_t|)\cdot|\nabla_j f_t|^2\right) \geq (1-\frac{1}{\kappa})^{-\kappa}\cdot\mu|\nabla_j f_t^{1-\frac{1}{\kappa}}|^\kappa \tag{8.1.9}$$

Thus , if $1-\frac{1}{\kappa}=\frac{1}{q}$, with $q \in (1,2)$, we have

$$\frac{d}{dt}S_t \leq -q^\kappa\cdot\sum_j \mu|\nabla_j f_t^{\frac{1}{q}}|^\kappa \tag{8.1.10}$$

Since $1 < q \leq \kappa$, we can use Hölder inequality for $l_p(\Lambda)$ norms to get

$$\frac{d}{dt}S_t \leq -q^\kappa|\Lambda|^{1-\frac{\kappa}{q}}\cdot\left(\sum_j \mu|\nabla_j f_t^{\frac{1}{q}}|^q\right)^{\frac{\kappa}{q}} \tag{8.1.11}$$

If now the following LS_q inequality is satisfied

$$\mathrm{Ent}_\mu(g) \leq q^q C\sum_j \mu|\nabla_j g^{\frac{1}{q}}|^q$$

then using (8.1.11), we obtain

$$\frac{d}{dt}S_t \leq -|\Lambda|^{1-\frac{\kappa}{q}}C^{-\frac{\kappa}{q}}\cdot S_t^{\frac{\kappa}{q}} \tag{8.1.12}$$

Solving this differential inequality one arrives at the following bound

$$S_t \leq \left(S_0^{-\frac{\kappa-q}{q}} + |\Lambda|^{1-\frac{\kappa}{q}}C^{-\frac{\kappa}{q}}(\frac{\kappa}{q}-1)\cdot t\right)^{-\frac{q}{\kappa-q}} \tag{8.1.13}$$

which implies (8.1.7).

Remark 8.6. We remark that as $q \longrightarrow 2$ the right hand side of (8.1.13) converges to the known exponential decay of entropy which is independent of Λ.

CHAPTER 9

Isoperimetry

Given a convex body K in $\mathbf{R}^n$ (a convex, compact, n-dimensional set with non-empty interior), denote by μ the corresponding normalised Lebesgue measure, that is, the uniform distribution on K. Thus, it has density

$$\frac{d\mu(x)}{dx} = \frac{1_K(x)}{\mathrm{vol}_n(K)}, \qquad x \in \mathbf{R}^n,$$

with respect to Lebesgue measure on $\mathbf{R}^n$, where we use 1_K to denote the indicator function of a set K. The canonical relative isoperimetric problem for K is to minimise the quantity $\mu(A + hB_2)$ provided that $\mu(A)$ and $h > 0$ are fixed (A is a Borel subset of $\mathbf{R}^n$ and B_2 is the unit Euclidean ball with centre at the origin). Another, formally weaker problem asks how to minimise the relative μ-perimeter

$$\mu_2^+(A) = \liminf_{h\downarrow 0} \frac{\mu(A+hB_2) - \mu(A)}{h}, \tag{9.0.1}$$

that is, how to find the so-called isoperimetric function for μ,

$$I_\mu(t) \;=\; \inf_{\mu(A)=t} \;\mu^+(A), \quad 0 < t < 1.$$

The problem makes sense for any probability measure μ on $\mathbf{R}^n$ and is of a large interest in Analysis and Probability.

Furthermore, we often obtain a different interesting problem when considering a different enlargement of sets by replacing the ball B_2 with other convex bodies B in $\mathbf{R}^n$ such as for example B_p, the unit balls with respect to ℓ^pmetric, ($1 \le p \le +\infty$). Appropriately, one should understand the notion of the perimeter and of the isoperimetric function.

Even in the traditional situation (the case of a convex body and of the Euclidean distance), the isoperimetric problem seems to be very far from being solved. One usually tries to study suitable families of convex bodies or some isoperimetric-type inequalities such as of the Cheeger-type

$$\mu_2^+(A) \ge c\,\min\{\mu(A), 1-\mu(A)\}, \quad A \subset \mathbf{R}^n \text{ Borel.} \tag{9.0.2}$$

The optimal value in such an inequality,

$$c(K) \;=\; \inf_{0<t\le\frac{1}{2}} \;\frac{I_\mu(t)}{t},$$

was introduced in 1969 by J. Cheeger [C] and since then it is called Cheeger's isoperimetric constant. Thus, it has to be estimated from below in terms of K or in terms of μ in the general measure case. In 1995, R. Kannan, L. Lovász, and

M. Simonovits [K-L-S] developed a localization technique which led to a number of interesting estimates for this quantity. In particular, they proved the bound

$$c(K) \geq \frac{c_0}{\int |x| \, d\mu(x)} \tag{9.0.3}$$

with some universal constant c_0. A different approach, based on the application of Prékopa-Leindler's theorem, was later suggested in [B3]. A weak point of this bound is that the dimension n is essentially involved in the right hand side of (1.3) and therefore one may not hope, for example, to reach on this way the concentration of measure phenomenon: in the simplest case of the unit Euclidean ball $K = B_2$, the Cheeger constant is of order $\sqrt{n}$, while the right hand side of (9.0.3) is close to a constant, as n tends to infinity. The same holds for all other convex bodies K which are brought in isotropic position.

Apparently this negative observation stimulated the authors of [K-L-S] to look for other better feeling the geometry of a convex body quantities . At the end of their paper, they introduced a quantity

$$\chi(K) = \int \chi_K(x) \, d\mu(x)$$

where $\chi_K(x)$ denotes the length of the longest interval in K with centre at $x \in K$, that is,

$$\chi_K(x) = \mathrm{diam}\,((K - x) \cap (x - K)). \tag{9.0.4}$$

Using it they derived the following isoperimetric inequality

$$\mu(A)(1 - \mu(A)) \leq \chi(K) \, \mu^+(A), \tag{9.0.5}$$

which is equivalent, up to a universal factor, to the bound

$$c(K) \geq \frac{1}{2\,\chi(K)}. \tag{9.0.6}$$

For the Euclidean balls, both sides of (9.0.6) are now of the same order. One should however emphasize that (9.0.6) remains inappropriate for some canonical cases. According to [K-L-S], (9.0.6) is not good for the regular simplex (and clearly for ℓ^1-ball $K = B_1$, as well). Another "negative" example is given by the unit ℓ^∞-ball $K = [-1, 1]^n$ when the Cheeger constant $c(K)$ is of order 1. In this case, $\chi_K^2(x) = \sum_i (1 - |x_i|)^2$, so $\chi(K)$ is of order $\frac{1}{\sqrt{n}}$.

The question of what is really essential to effectively estimate the isoperimetric function I_μ and, in particular, the Cheeger constant $c(K)$ is still open. There is however a striking hypothesis suggested in the same paper [K-L-S] which says the following: if we consider the inequality (9.0.2) in the class of all half-spaces A of $\mathbf{R}^n$, the corresponding optimal constant c should be equivalent to $c(K)$ up to universal factors. The same can be conjectured for general log-concave probability measures μ on $\mathbf{R}^n$. For example, the hypothesis is true in the class of product log-concave measures; this was shown in [B-H1] with the help of a suitable induction argument going back to [B2].

Although, too little is known about sharp bounds in the general case, the K-L-S theorem (9.0.5)-(9.0.6) seems to be well adapted to study uniformly convex bodies and uniformly convex probability measures. The isoperimetric problem for such measures was first investigated in the 1987 paper by M. Gromov and V. D. Milman [G-M]. They introduced a special kind of localization allowing to work on the sphere of a uniformly convex Banach space; see also [A]. For various localization arguments, we can refer the reader to the recent works [F-G] and [N-S-V].

Let us return to the isoperimetric inequality (9.0.2) and write it as the relation

$$I_\mu(t) \geq c \min\{t, 1-t\}, \quad 0 < t < 1.$$

How sharp is it in general ? It is well known that, in the body case, the isoperimetric function $I_\mu(t)$ for small t resembles very much the function $c\,t^{(n-1)/n}$. Therefore one also considers (especially in Riemannian Geometry, cf. [Cr], [Li]) the so-called Sobolev constant

$$s_n(K) \;=\; \inf_{0<t\leq\frac{1}{2}} \frac{I_\mu(t)}{t^{\frac{n-1}{n}}}.$$

It turns out that this more sensitive quantity (in comparison with $c(K)$) can also be related to the geometric characteristic $\chi(K)$, and we have the following refinement of (9.0.6):

Theorem 9.1. *For any convex body K in $\mathbf{R}^n$,*

$$s_n(K) \geq \frac{1}{C\chi(K)}, \tag{9.0.7}$$

where $C > 0$ is a numerical constant.

The proof uses a localization argument of [K-L-S] which we discuss in the next section. To be more precise, assume $n \geq 2$ and define another quantity

$$\chi_n(K) = \left(\int \chi_K(x)^{\frac{n}{n-1}}\, d\mu(x)\right)^{\frac{n-1}{n}} \;=\; \|\,\chi_K\,\|_{L^{\frac{n}{n-1}}(K,\mu)}.$$

With this quantity we have the following result.

Theorem 9.2. *Given a convex body K in $\mathbf{R}^n$ with the normalised Lebesgue measure μ on it, for all Borel $A \subset \mathbf{R}^n$,*

$$\mu(A)^{\frac{n-1}{n}}(1-\mu(A)) \leq 4\chi_n(K)\,\mu^+(A). \tag{9.0.8}$$

In particular,

$$s_n(K) \geq \frac{1}{8\,\chi_n(K)}. \tag{9.0.9}$$

These estimates remain to hold with respect to any norm on $\mathbf{R}^n$ with the corresponding notion of the diameter in (9.0.4) *and of the perimeter in* (9.0.1).

As we will see, in the Euclidean case, $\chi_n(K) \leq C\chi(K)$, for some universal constant C, so Theorem 9.2 is more general. We will illustrate the bound (9.0.9) on the example of uniformly convex bodies in $\mathbf{R}^n$.

Let us also mention a well known relationship of the Sobolev constants with Sobolev-type inequalities (which is due to V. G. Maz'ja [M1], cf. also [M2], [B-Z]). The combination of the coarea formula with (9.0.8) allows one easily to deduce:

Corollary 9.3. *For any convex body K in $\mathbf{R}^n$, and every smooth function f on $\mathbf{R}^n$ with $\int_K f(x)\,dx = 0$, with respect to any norm $\|\cdot\|$ on $\mathbf{R}^n$,*

$$\left(\int_K |f(x)|^{\frac{n}{n-1}}\, d\mu(x)\right)^{\frac{n-1}{n}} \;\leq\; C\chi_n(K) \int_K \|\nabla f(x)\|_*\, d\mu(x), \tag{9.0.10}$$

where $\| \cdot \|_*$ *is the dual norm and* C *is a numerical constant.*

CHAPTER 10

The localization argument

The localization argument of [K-L-S] can be applied to a variety of multidimensional geometric inequalities. In particular, it can be applied to (9.0.8). With the following lemma due to L. Lovász and M. Simonovits [L-S] it can be then reduced to a certain analytic problem in dimension one.

Lemma 10.1. *Let u and v be two lower semi-continuous, integrable functions on $\mathbf{R}^n$ such that*

$$\int_{\mathbf{R}^n} u(x)\,dx > 0, \qquad \int_{\mathbf{R}^n} v(x)\,dx > 0.$$

Then for some points $a, b \in \mathbf{R}^n$ and some affine function $\ell : (0,1) \to (0,+\infty)$,

$$\int_0^1 u((1-t)a+tb)\,\ell(t)^{n-1}\,dt > 0, \qquad \int_0^1 v((1-t)a+tb)\,\ell(t)^{n-1}\,dt > 0.$$

Recall that a function is lower semi-continuous, if it can be represented as a pointwise limit of an increasing sequence of continuous functions. For example, the indicator function of an open set is lower semi-continuous. Lemma 10.1 implies:

Lemma 10.2. *Let $\alpha, \beta > 0$ and let u_1, u_2, u_3, u_4 be continuous, non-negative functions defined on a convex body K in $\mathbf{R}^n$ and such that, for all $a, b \in K$ and for any affine function $\ell : (0,1) \to (0,+\infty)$,*

$$\begin{aligned} &\left(\int_0^1 u_1((1-t)a+tb)\,\ell(t)^{n-1}\,dt\right)^\alpha \left(\int_0^1 u_2((1-t)a+tb)\,\ell(t)^{n-1}\,dt\right)^\beta \\ &\le \left(\int_0^1 u_3((1-t)a+tb)\,\ell(t)^{n-1}\,dt\right)^\alpha \left(\int_0^1 u_4((1-t)a+tb)\,\ell(t)^{n-1}\,dt\right)^\beta. \end{aligned} \tag{10.0.1}$$

Then,

$$\left(\int_K u_1(x)\,dx\right)^\alpha \left(\int_K u_2(x)\,dx\right)^\beta \le \left(\int_K u_3(x)\right)^\alpha \left(\int_K u_4(x)\,dx\right)^\beta. \tag{10.0.2}$$

Lemma 10.2 is formulated in [K-L-S] in a slightly different way in terms of the so-called exponential needles. The corresponding arguments are as follows.

Without loss of generality, let the functions u_j be strictly positive so that all the considered integrals (including one dimensional) do not vanish. By considering an ε-interior of K and then letting $\varepsilon \downarrow 0$, we may also assume that K is open and bounded.

Now, assuming the contrary to (10.0.2), one can take a number $A > 0$ such that

$$\left(\frac{\int_K u_1}{\int_K u_3}\right)^\alpha > A > \left(\frac{\int_K u_4}{\int_K u_2}\right)^\beta,$$

or equivalently,

$$\int_{\mathbf{R}^n} (u_1 - A^{1/\alpha} u_3)\, 1_K > 0, \quad \int_{\mathbf{R}^n} (A^{1/\beta} u_2 - u_4)\, 1_K > 0.$$

The functions under the integral sign are lower semi-continuous, so, by Lemma 10.1, one can find a segment $(a,b) \subset \mathbf{R}^n$ and a positive affine function ℓ on (a,b) such that

$$\int_{(a,b)} (u_1 - A^{1/\alpha} u_3)\, \ell^{n-1}\, 1_K > 0, \quad \int_{(a,b)} (A^{1/\beta} u_2 - u_4)\, \ell^{n-1}\, 1_K > 0,$$

where the integrals are understood with respect to Lebesgue measure on (a,b). In terms of the interval $(a',b') = (a,b) \cap K \subset K$, we have

$$\int_{(a',b')} (u_1 - A^{1/\alpha} u_3)\, \ell^{n-1} > 0, \quad \int_{(a',b')} (A^{1/\beta} u_2 - u_4)\, \ell^{n-1} > 0.$$

Since again these two inequalities may be written as

$$\left(\frac{\int_{(a',b')} u_1\, \ell^{n-1}}{\int_{(a',b')} u_3\, \ell^{n-1}}\right)^\alpha > A > \left(\frac{\int_{(a',b')} u_4\, \ell^{n-1}}{\int_{(a',b')} u_2\, \ell^{n-1}}\right)^\beta,$$

we obtain a contradiction with (10.0.1).

Clearly (using suitable approximation), Lemma 10.2 remains to hold for many discontinuous functions involving, for example, indicator functions of "regular" subsets of $\mathbf{R}^n$. To derive (9.0.5), Kannan, Lovász, and Simonovits considered the inequality of the form

$$\mu(A)\, \mu(B) \leq \int_K \chi_K(x)\, d\mu(x)\, \frac{\mu(C)}{h}, \quad h > 0, \tag{10.0.3}$$

where A, B, C is an arbitrary partition of K into non-empty "regular" subsets such that $dist(A,B) = h$. Since it is of the form (10.0.2) with

$$\alpha = \beta = 1, \quad u_1 = 1_A, \quad u_2 = 1_B, \quad u_3 = \chi_K, \quad u_4 = \frac{1_C}{h},$$

we are reduced to the corresponding inequality (10.0.1) for dimension one. On the other hand, fixing a set A, taking $B = K \setminus (A + hB_2)$, $C = K \setminus (A \cup B)$, and letting h tend to zero, the inequality (10.0.3) will turn unto the bound (9.0.5) for the Cheeger constant.

Similarly and better for our purposes, one may study an inequality of the form

$$c\, \mu(A)^{\frac{n-1}{n}}\, \mu(B) \leq \left(\int_K \chi_K(x)^{\frac{n}{n-1}}\, d\mu(x)\right)^{\frac{n-1}{n}} \frac{\mu(C)}{h}, \tag{10.0.4}$$

which for small h becomes the desired inequality (9.0.8) of Theorem 9.2 (the constant $c > 0$ has to be precised later on the basis of a certain one dimensional problem). Again, we are in position to apply Lemma 10.2 with

$$\alpha = \frac{n-1}{n}, \quad \beta = 1, \quad u_1 = 1_A, \quad u_2 = c\, 1_B, \quad u_3 = \chi_K^{\frac{n}{n-1}}, \quad u_4 = \frac{1_C}{h},$$

so we are reduced to the particular case of (10.0.1) which can be written as follows

$$c\,\mu_\ell(A)^{\frac{n-1}{n}}\,\mu_\ell(B) \le \left(\int_{(a,b)} \chi_K(x)^{\frac{n}{n-1}}\,d\mu_\ell(x)\right)^{\frac{n-1}{n}} \frac{\mu_\ell(C)}{h}. \tag{10.0.5}$$

Here, ℓ is a positive, affine function defined on a segment $(a,b) \subset K$, and μ_ℓ denotes the measure on (a,b) with density ℓ^{n-1} with respect to Lebesgue measure on (a,b). Recall that, (because the distance is determined by the norm $\|\cdot\|$ in $\mathbf{R}^n$), we have

$$h = \mathrm{dist}(A,B) = \inf\{\,\|a-b\| : a \in A,\ b \in B\},$$

$$\chi_K(x) = 2\,\max\{\,\|u\| : x \pm u \in K,\ u \in \mathbf{R}^n\}.$$

The inequality (10.0.5) will formally become stronger, if the sets A, B, C are replaced respectively by $A \cap (a,b)$, $B \cap (a,b)$, $C \cap (a,b)$: in this case the distance h will only become larger. Therefore, in (10.0.5) it suffices to consider partitions of (a,b). Moreover, the inequality will formally be strengthened if we replace χ_K by the smaller function

$$\chi_{(a,b)}(x) = 2\,\sup\{\,\|u\| : x \pm u \in (a,b),\ u \in \mathbf{R}^n\}, \quad x \in (a,b).$$

Thus, (10.0.5) is reduced to the form

$$c\,\mu_\ell(A)^{\frac{n-1}{n}}\,\mu_\ell(B) \le \left(\int_{(a,b)} \chi_{(a,b)}(x)^{\frac{n}{n-1}}\,d\mu_\ell(x)\right)^{\frac{n-1}{n}} \frac{\mu_\ell(C)}{h}, \tag{10.0.6}$$

where A, B, C is an arbitrary "regular" partition of the segment (a,b) such that $h = dist(A,B)$.

This problem, where the set K has been eliminated, is already a purely one dimensional problem. For convenience, it may be formulated on the segment $(0,1)$ of the real line using the parametrisation of (a,b) by the map $x_t = a + t(b-a)$, $0 < t < 1$. Then, $\|x_t - x_s\| = \|b-a\|\,|t-s|$, so

$$h = \|b-a\|\,\mathrm{dist}(A',B'),$$

where $A' = \{t \in (0,1) : x_t \in A\}$, $B' = \{t \in (0,1) : x_t \in B\}$, and where we use the usual distance on the real line. On the hand, the vector u appearing in the definition of $\chi_{(a,b)}$ has to be of the form $s(b-a)$, for some $s \in \mathbf{R}$. Therefore, as easy to see,

$$x_t + u \in (a,b) \text{ and } x_t - u \in (a,b) \iff |s| < \min\{t, 1-t\}.$$

Consequently, $\chi_{(a,b)}(x_t) = \|b-a\|\,\min\{t,1-t\}$. Introducing $C' = \{t \in (0,1) : x_t \in C\}$ and the measure μ'_ℓ on $(0,1)$ – pre-image of μ_ℓ under the parametrisation map, we can thus rewrite (10.0.6) as

$$c\,\mu'_\ell(A')^{\frac{n-1}{n}}\,\mu'_\ell(B') \le \left(\int_0^1 (\min\{t,1-t\})^{\frac{n}{n-1}}\,d\mu'_\ell(t)\right)^{\frac{n-1}{n}} \frac{\mu_\ell(C')}{\mathrm{dist}(A',B')}.$$

Finally, note that, up to a numerical constant, the measure μ'_ℓ has density $\ell(x_t)^{n-1}$ and that $\ell(x_t)$ is an affine and positive function in $t \in (0,1)$. Thus we conclude with:

Corollary 10.3. *Given $c > 0$ and $h > 0$, the multidimensional inequality (10.0.4) is equivalent to the property that*

$$c\,\mu_\ell(A)^{\frac{n-1}{n}}\,\mu_\ell(B) \le \left(\int_0^1 (\min\{t, 1-t\})^{\frac{n}{n-1}}\,d\mu_\ell(t)\right)^{\frac{n-1}{n}} \frac{\mu_\ell(C)}{h}, \tag{10.0.7}$$

where A, B, C is a partition of $(0,1)$ with $dist(A,B) = h$, where ℓ is a affine, positive function on $(0,1)$, and μ_ℓ is a measure on $(0,1)$ with density ℓ^{n-1}.

The above argument showed that (10.0.7) implies (10.0.4). In turn, choosing infinitesimally small truncated cones K in (10.0.4), one obtains in the limit (10.0.7).

Note that the inequality (10.0.7) is homogeneous with respect to ℓ, so we may always assume that μ_ℓ is a probability measure, that is, $\int_0^1 \ell(t)^{n-1}\,dt = 1$.

CHAPTER 11

Infinitesimal version

To prove Theorem 9.2, we are now faced with the one dimensional problem (10.0.7) where finding an optimal (or some) constant c does seem a much simpler, although not yet immediate, task. Recall that the inequality (10.0.4) yields the desired result by letting $h \to 0$. Similarly, we would obtain a simpler one dimensional inequality by letting $h \to 0$ in (10.0.7). This would potentially simplify our task, but then, can we return back? That is, will we obtain an equivalent inequality? An affirmative answer is given in the following:

Proposition 11.1. *In the setting of an arbitrary probability metric space* (M, ρ, μ), *for any* $\alpha \in [\frac{1}{2}, 1]$ *and* $c > 0$, *the inequality of the form*

$$c\,\mu(A)^{\alpha}\,\mu(B) \leq \frac{\mu(C)}{h}, \tag{11.0.1}$$

where A, B, C *is an arbitrary partition of* M *into Borel measurable subsets such that* $h = dist(A, B)$, *is equivalent to its infinitesimal version obtained by letting* $h \to 0$,

$$c\,\mu(A)^{\alpha}\,(1 - \mu(A)) \leq \mu^{+}(A), \quad A \subset M \text{ Borel.} \tag{11.0.2}$$

In the general case, the perimeter is defined similarly to (9.0.1) as

$$\mu^{+}(A) = \liminf_{h \downarrow 0} \frac{\mu(A^h) - \mu(A)}{h},$$

where $A^h = \{x \in M : \rho(x, a) < h,\ for\ some\ a \in A\}$ is the open h-neighbourhood of A with respect to ρ. With this notations, and since the right hand side of (11.0.1) is continuous in h, one can put in (11.0.1) $B = M \setminus A^h$, $C = A^h \setminus A$, and then the inequality itself will take the form

$$\mu(A^h) \geq \frac{\mu(A) + c \cdot h\,\mu(A)^{\alpha}}{1 + c \cdot h\,\mu(A)^{\alpha}}, \quad A \subset M \text{ Borel.}$$

In the sequel (throughout this section), since the parameter c may be absorbed by the variable h, we assume $c = 1$, so that (11.0.1) has the form

$$\mu(A^h) \geq L_h(\mu(A)), \quad h > 0, \tag{11.0.3}$$

for the family

$$L_h(p) = \frac{p + hp^{\alpha}}{1 + hp^{\alpha}}, \quad 0 < p < 1.$$

In turn, write (11.0.2) as

$$\mu^{+}(A) \geq I(\mu(A)), \quad 0 < \mu(A) < 1, \tag{11.0.4}$$

for

$$I(p) = p^\alpha q, \quad 0 < p < 1, \; q = 1 - p.$$

It is well known, (cf. e.g. [B-H2], [L]), that given a continuous non-negative function I on (0,1), the isoperimetric inequality (11.0.4) can be "integrated over h" and equivalently be written as

$$\mu(A^h) \geq R_h(\mu(A)), \quad h > 0, \; \mu(A) > 0, \tag{11.0.5}$$

for a suitable semi-group $\{R_h\}_{h \geq 0}$. Namely, one can associate with I a distribution function on the real line, $F = F(x)$, $x \in \mathbf{R}$, defined via its inverse by

$$F^{-1}(p) = \int_{1/2}^{p} \frac{dt}{I(t)}, \quad 0 < p < 1.$$

It has median at zero and, as a measure, F is concentrated on the interval $(a, b) \subset \mathbf{R}$, finite or not, with

$$a = -\int_0^{1/2} \frac{dt}{I(t)}, \quad b = \int_{1/2}^{1} \frac{dt}{I(t)}.$$

In our concrete case, a is finite (if $\alpha < 1$), while $b = +\infty$; although this will not matter later. What is important for us is that one can take in (11.0.5)

$$R_h(p) = F(F^{-1}(p) + h), \quad 0 \leq p \leq 1. \tag{11.0.6}$$

Here, by continuity, $R_h(0) = F(a + h)$, $R_h(1) = 1$.

Recalling (11.0.3), and in view of the equivalence of (11.0.4) and (11.0.5), Proposition 11.1 will be implied by:

Proposition 11.2. *For $I(p) = p^\alpha q$ with $\frac{1}{2} \leq \alpha \leq 1$, we have $R_h(p) \geq L_h(p)$, for all $h > 0$ and $p \in [0, 1]$.*

For the convenience, we divide the proof into several steps. First we state one crucial property of the family L_h. Clearly, these functions bijectively act from $[0, 1]$ onto itself.

Lemma 11.3. *For all $h_1, h_2 \geq 0$ and $p \in [0, 1]$,*

$$L_{h_1 + h_2}(p) \leq L_{h_1}(L_{h_2}(p)). \tag{11.0.7}$$

Proof. Indeed, using $L_h(p) = 1 - \frac{q}{1 + hp^\alpha}$, the property (11.0.7) is just

$$\frac{1 - L_{h_2}(p)}{1 + h_1 L_{h_2}(p)^\alpha} \leq \frac{q}{1 + (h_1 + h_2)p^\alpha},$$

that is,

$$(1 - L_{h_2}(p))(1 + (h_1 + h_2)p^\alpha) \leq q\,(1 + h_1 L_{h_2}(p)^\alpha).$$

This inequality is affine with respect to h_1, so the cases $h_1 = 0$ and $h_1 = +\infty$ have to be considered, only. When $h_1 = 0$, we obtain equality (by the definition of L_h). Letting $h_1 \to +\infty$, we are reduced to

$$(1 - L_{h_2}(p))p^\alpha \leq q L_{h_2}(p)^\alpha \iff \frac{p^\alpha q}{1 + h_2 p^\alpha} \leq q L_{h_2}(p)^\alpha \iff \frac{p}{(1 + h_2 p^\alpha)^{1/\alpha}} \leq L_{h_2}(p).$$

Replacing h_2 by h and using the definition of L_h, we arrive at

$$p\,(1 + hp^\alpha)^{\frac{1}{\alpha} - 1} \leq p + hp^\alpha. \tag{11.0.8}$$

Note that, due to the assumption $\alpha \in [\frac{1}{2}, 1]$, the left hand side represents a non-decreasing, concave function in $h \geq 0$, while the right hand side is affine. Hence, it suffices to consider the behaviour of both sides near zero. When $h = 0$, (11.0.8) turns into equality, while the corresponding derivatives at $h = 0$ turn the inequality into $p\,(\frac{1}{\alpha} - 1) \leq 1$ which is true. Thus, the property (11.0.7) is verified.

In contrast with (11.0.7), it follows immediately from the definition (11.0.6) that:

Lemma 11.4. *For all $h_1, h_2 \geq 0$ and $p \in [0,1]$, $R_{h_1+h_2}(p) = R_{h_1}(R_{h_2}(p))$.*

Lemma 11.5. *For $I(p) = p^\alpha q$ with $\frac{1}{2} \leq \alpha \leq 1$, there exist $0 < p_0 < p_1 < 1$ and $h_0 > 0$ such that*

$$R_h(p) \geq L_h(p), \tag{11.0.9}$$

whenever $h \in [0, h_0]$ and $p \in [0, p_0] \cup [p_1, 1]$.

Proof. We may assume $\frac{1}{2} < \alpha < 1$ to simplify the argument in some details. The function $I(p) = p^\alpha q$ is concave on (0,1), so F has log-concave density $f(x) = F'(x)$ on the supporting interval (a, b). Recall that, according to the definition of F,

$$f(F^{-1}(p)) = I(p) = p^\alpha q, \quad 0 < p < 1.$$

Clearly, the density f is smooth on (a, b). Formally differentiating the above equality, we get

$$f'(F^{-1}(p)) = I(p)I'(p) = p^{2\alpha-1} q\,(\alpha q - p), \quad 0 < p < 1. \tag{11.0.10}$$

Further differentiation of $J(p) = f'(F^{-1}(p))$ gives

$$J'(p) = p^{2\alpha-2}\left((\alpha+1)(2\alpha+1)p^2 - 2\alpha(2\alpha+1)p + \alpha(2\alpha-1)\right).$$

Thus, for small p, $J'(p)$ behaves like $\alpha(2\alpha-1)p^{2\alpha-2} > 0$, so

$$J'(0+) = +\infty, \quad J'(1-) = 1.$$

Hence, $J(p)$ increases near $p = 0$ and near $p = 1$, and the same is true for the function $f'(x)$ near $x = a$ and $x = +\infty$.

For a fixed $p \in (0,1)$, write Taylor's expansion for $R_h(p) - F(F^{-1}(p) + h)$ as a function of h up to the quadratic term:

$$R_h(p) = p + I(p)h + f'(F^{-1}(p) + w)\,\frac{h^2}{2}, \quad 0 < w < h.$$

If p is sufficiently close to 1, or if p and h are sufficiently close to 0, by the above mentioned monotonicity,

$$R_h(p) \geq p + I(p)h + f'(F^{-1}(p))\,\frac{h^2}{2}.$$

Therefore, by (11.0.10) and since $L_h(p) = 1 - \frac{q}{1+hp^\alpha}$, (11.0.9) would be implied by

$$p + I(p)h + I(p)\,I'(p)\,\frac{h^2}{2} \geq 1 - \frac{q}{1+hp^\alpha}.$$

Subtracting p and dividing by $I(p)h$, this inequality becomes $1 + I'(p)\,\frac{h}{2} \geq \frac{1}{1+hp^\alpha}$, or, $-\frac{1}{2}\,I'(p) \leq \frac{p^\alpha}{1+hp^\alpha}$. For small p, this bound is immediate, since I increases near zero. To treat the other case, rewrite the bound as $\frac{1}{2}\,(p - \alpha q) \leq \frac{p}{1+hp^\alpha}$. This inequality is readily fulfilled for sufficiently small q and h (independently of each other). Hence Lemma 11.5 follows.

Proof of Proposition 11.2. Given $t \in (0,1)$, define

$$\Delta = \{h \geq 0 : R_h(p) \geq L_{th}(p), \text{ for all } p \in [0,1]\}.$$

If we show that $\Delta = [0, +\infty)$, then letting $t \to 1$ will give the desired inequality (11.0.9). For any $h_1, h_2 \in \Delta$ and $p \in [0,1]$, by Lemmas 11.3 and 11.4,

$$R_{h_1+h_2}(p) = R_{h_1}(R_{h_2}(p)) \geq L_{th_1}(L_{th_2}(p)) \geq L_{t(h_1+h_2)}(p).$$

Hence, $h_1 + h_2 \in \Delta$, so $\Delta + \Delta = \Delta$. Therefore, it suffices to show that Δ contains all sufficiently small $h > 0$. By Lemma 11.5, in the inequality $R_h(p) \geq L_{th}(p)$, it remains to consider the values $p_0 \leq p \leq p_1$, only. Writing the Taylor's expansion over $h \downarrow 0$ for the functions $R_h(p)$ and $L_{th}(p)$ we have:

$$R_h(p) = p + I(p)h + O(h^2), \quad L_h(p) = p + tI(p)h + O(h^2). \tag{11.0.11}$$

Here both constants in $O(h^2)$ are uniform in p, since the corresponding second derivatives are bounded. Indeed, by (11.0.10), f' is bounded, so is $\frac{d^2 R_h(p)}{dh^2} = f'(F^{-1}(p) + h)$. In addition,

$$\frac{d^2 L_h(p)}{dh^2} = -\frac{2p^{2\alpha} q}{(1 + hp^\alpha)^3}$$

is evidently bounded. But $I(p)$ is separated from zero on $[p_0.p_1]$, so, for small $h > 0$, the inequality $R_h(p) \geq L_{th}(p)$ immediately follows from (11.0.11).

CHAPTER 12

Proof of Theorem 9.2

Now we can return to the one dimensional inequality (10.0.7) and apply Proposition 11.1 to rewrite (10.0.7) in the infinitesimal form as

$$(12.0.1)\qquad c\,\mu_\ell(A)^\alpha\,(1-\mu_\ell(A)) \le \left(\int_0^1 (\min\{t,1-t\})^{1/\alpha}\,d\mu_\ell(t)\right)^\alpha \mu_\ell^+(A).$$

Here A is an arbitrary Borel subset of (0,1) and $\alpha = \frac{n-1}{n}$. Note that $\frac{1}{2} \le \alpha < 1$, for any integer $n \ge 2$. Since $L^{1/\alpha}$-norm majorizes L^1-norm, (12.1) can be strengthened as

$$(12.0.2)\qquad c\,\mu_\ell(A)^\alpha\,(1-\mu_\ell(A)) \le \int_0^1 \min\{t,1-t\}\,d\mu_\ell(t)\ \mu_\ell^+(A)$$

(remaining equivalent up to a numerical factor). Note that, in terms of the distribution function $F(t) = \mu_\ell[0,t]$, $0 \le t \le 1$, the above integral has a simple representation

$$(12.0.3)\qquad \int_0^1 \min\{t,1-t\}\,d\mu_\ell(t) = \int_0^{1/2} \left(F\left(t+\frac{1}{2}\right) - F(t)\right) dt.$$

Thus, we are dealing in (12.0.2) with an inequality on the real line of the form

$$(12.0.4)\qquad \bar{c}\,\lambda(A)^\alpha\,(1-\lambda(A)) \le \lambda^+(A)$$

for special probability measures $\lambda = \mu_\ell$ (the new constant $\bar{c}$ temporarily absorbs the other factor in (12.0.2) independent of A). Since the non-negative functions of the form ℓ^{n-1} on [0,1] extended to the whole line by zero outside $[0,1]$ are log-concave, the measures μ_ℓ are log-concave in the sense of Prékopa (cf. [Pr], [Bor2]). The isoperimetric problem for any log-concave measure λ on R has a simple solution: the perimeter $\lambda^+(A)$ attains minimum within all Borel measurable sets A with prescribed value of $\lambda(A)$ either for a half-axis $A = (-\infty, x]$, or for $A = [x, +\infty)$; see [B1] for details. Equivalently, in terms of the distribution function $F(x) = \lambda(-\infty, x]$ and its density f, we have a simple representation for the isoperimetric function of such a measure,

$$(12.0.5)\qquad I_\lambda(p) = \min\{f(F^{-1}(p)), f(F^{-1}(q))\}, \quad 0 < p < 1,\ q = 1-p,$$

where F^{-1} is the corresponding inverse function. Therefore, the inequality (12.0.4) is needed to verify on half-axes, only, and the optimal value of $\bar{c}$ comes from

$$(12.0.6)\qquad \bar{c}\,p^\alpha q \le I_\lambda(p)$$

with the right hand side described in (12.0.5).

So, let us find an analytic expression for the isoperimetric function in case $\lambda = \mu_\ell$. Since $I_\lambda(p)$ is always symmetric about the point $p = \frac{1}{2}$, and since the integral in (12.0.2) is invariant under the transformation $t \to 1-t$, we may assume that $\ell(t)$ is non-decreasing in t and thus has the form

$$\ell(t) = \text{const}\,(r+t)$$

with necessarily $r \geq 0$. In view of the condition $\mu_\ell[0,1] = 1$, the distribution and density functions of this measure are given by

$$F_r(t) = \frac{(r+t)^n - r^n}{(r+1)^n - r^n}, \quad f_r(t) = \frac{n}{(r+1)^n - r^n}\,(r+t)^{n-1} \quad (0 < t < 1), \tag{12.0.7}$$

where r is a non-negative parameter. From this, as it is easy to see,

$$f_r(F_r^{-1}(p)) = \frac{n}{(r+1)^n - r^n}\,(p(r+1)^n + qr^n)^\alpha.$$

Given $x, y \geq 0$, the quantity $\frac{(xp+yq)^\alpha}{p^\alpha q} = (x + \frac{y}{q})^\alpha \frac{1}{q}$ attains minimum for $q = 1$, when it equals $x + y$. In particular,

$$\inf_{0<p<1} \frac{(p(r+1)^n + qr^n)^\alpha}{p^\alpha q} = \inf_{0<p<1} \frac{(q(r+1)^n + pr^n)^\alpha}{p^\alpha q} = ((r+1)^n + r^n)^\alpha.$$

Hence, by (12.0.5), the optimal constant in (12.0.6) for the isoperimetric function I_λ of μ_ℓ satisfies

$$\bar{c} = n\,\frac{((r+1)^n + r^n)^\alpha}{(r+1)^n - r^n} \geq n\,\frac{(r+1)^{n-1}}{(r+1)^n - r^n}. \tag{12.0.8}$$

Now, according to (12.0.7),

$$\int_0^{1/2} \left(F_r\left(t + \frac{1}{2}\right) - F_r(t)\right) dt = \frac{1}{n+1}\,\frac{(r+1)^{n+1} - 2(r+\frac{1}{2})^{n+1} + r^{n+1}}{(r+1)^n - r^n}. \tag{12.0.9}$$

Recalling (12.0.2)-(12.0.3) and applying (12.0.8)-(12.0.9), we may summarize:

Lemma 12.1. *The inequality* (12.0.1) *for the measure* μ_ℓ *defined in* (12.0.7) *with parameter* $r \geq 0$ *holds true with*

$$c = \frac{n}{n+1}\,\frac{(r+1)^{n-1}\left((r+1)^{n+1} - 2\,(r+\frac{1}{2})^{n+1} + r^{n+1}\right)}{((r+1)^n - r^n)^2}. \tag{12.0.10}$$

At last, we are ready to complete the proof of Theorem 9.2.

Proof of Theorem 9.2. It remains to estimate from below by a numerical constant the right hand side of (12.0.10). So, let us look at the inequality of the form

$$(r+1)^{n-1}\left((r+1)^{n+1} - 2\left(r + \frac{1}{2}\right)^{n+1} + r^{n+1}\right) \geq D_n\,((r+1)^n - r^n)^2. \tag{12.0.11}$$

Note that both sides represent certain polynomials of degree $2(n-1)$. More precisely, according to Newton's binomial formula,

$$((r+1)^n - r^n)^2 = \sum_{i=1}^n \sum_{j=1}^n C_n^i C_n^j\, r^{2n-(i+j)},$$

where $C_n^i = \frac{n!}{i!(n-i)!}$ are usual binomial coefficients. Writing $(r+1)^{n-1} = \sum_{i=1}^n C_{n-1}^{i-1}\, r^{n-i}$ and noting that

$$(r+1)^{n+1} - 2\left(r+\frac{1}{2}\right)^{n+1} + r^{n+1} = \sum_{j=1}^n C_{n+1}^{j+1}\left(1-\frac{1}{2^j}\right) r^{n-j} \geq \frac{1}{2}\sum_{j=1}^n C_{n+1}^{j+1}\, r^{n-j}.$$

we see that (12.0.11) is implied by

$$\sum_{i=1}^n\sum_{j=1}^n C_{n-1}^{i-1} C_{n+1}^{j+1}\, r^{2n-(i+j)} \geq 2D_n\left(\sum_{i=1}^n\sum_{j=1}^n C_n^i C_n^j\, r^{2n-(i+j)}\right). \tag{12.0.12}$$

Let us compare here the coefficients $A_{ij} = C_{n-1}^{i-1} C_{n+1}^{j+1}$ with $B_{ij} = C_n^i C_n^j$. We have in general $A_{ij} = \frac{i}{j+1}\,\frac{n+1}{n}\, B_{ij}$, so

$$A_{ii} = \frac{i}{i+1}\,\frac{n+1}{n}\, B_{ii} \geq \frac{1}{2}\,\frac{n+1}{n}\, B_{ii}, \quad 1 \leq i \leq n.$$

In view of $B_{ij} = B_{ji}$, for the case $1 \leq i \neq j \leq n$ we also have

$$A_{ij} + A_{ji} = \left(\frac{i}{j+1} + \frac{j}{i+1}\right)\frac{n+1}{n}\,\frac{B_{ij}+B_{ji}}{2} \geq \frac{1}{2}\,\frac{n+1}{n}\,(B_{ij}+B_{ji}),$$

because one of the fractions $\frac{i}{j+1}$, $\frac{j}{i+1}$ is greater or equal to 1. This shows that (12.0.12) holds true with $2D_n = \frac{1}{2}\,\frac{n+1}{n}$. Therefore, by Lemma 12.1, we may take $c = \frac{1}{4}$ in (12.0.1) and Theorem 9.2 is now proved.

CHAPTER 13

Euclidean distance (proof of Theorem 9.1)

Since $\frac{n}{n-1} \leq 2$, in order to derive Theorem 9.1 from Theorem 9.2, it will be sufficient to establish a Khichine-type inequality for the functions χ_K. As we will see, it holds true with respect to any log-concave measure.

Proposition 13.1. *For any convex body K in $\mathbf{R}^n$ and every log-concave probability measure λ concentrated on K,*

$$\left(\int \chi_K^2 \, d\lambda\right)^{1/2} \leq C \int \chi_K \, d\lambda, \tag{13.0.1}$$

where C is a numerical constant.

Proof. Log-concavity of a measure λ on $\mathbf{R}^n$ means (cf. [Bor2]) that, for all $t, s > 0$ with $t + s = 1$ and for all non-empty Borel subsets A, B, C of R^n with $tA + sB \subset C$,

$$\lambda(C) \geq \lambda(A)^t \lambda(B)^s.$$

The uniform distribution $\lambda = \mu$ on K possesses this property due to Brunn-Minkowski inequality (cf. e.g. [B-Z]).

By (9.0.4) for the distance given by the Euclidean norm $|\bullet|$ in $\mathbf{R}^n$, for any $x \in K$ one has

$$\chi_K(x) = 2 \max\{|u| : x \pm u \in K, \ u \in \mathbf{R}^n\}.$$

For given $x, y \in K$, let the vectors $u, v \in \mathbf{R}^n$ be the maximizers in this definition. Since for all $t, s > 0$ with $t + s = 1$, necessarily $tx + sy \pm (tu \pm sv) \in K$, we get

$$\chi_K^2(tx + sy) > 4\,|tu \pm sv|^2 = 4\left(t^2|u|^2 \pm 2ts\,\langle u, v\rangle + s^2|v|^2\right).$$

Choosing appropriately the sign on the right hand side, we obtain

$$\chi_K^2(tx + sy) \geq t^2 \chi_K^2(x) + s^2 \chi_K^2(y), \quad x, y \in K, \ t, s > 0, t + s = 1. \tag{13.0.2}$$

Let $D = \max_{x \in K} \chi_K^2(x)$ (clearly, the function χ_K is continuous, so the maximum is attained at some point), and let $\xi = \chi_K^2$ be considered as a random variable on the probability space (K, λ) so that

$$\begin{aligned} \int \chi_K^2 \, d\lambda &= \mathbf{E}\,\xi = \int_0^D \lambda\{\xi \geq a\} \, da, \\ \int \chi_K \, d\lambda &= \mathbf{E}\,\xi^{1/2} = \int_0^D \lambda\{\xi \geq a\} \, da^{1/2}. \end{aligned}$$

For any $a \in [0, D]$, define the set $A(a) = \{x \in K : \xi(x) \geq a\}$. By (13.0.2), for all $a, b \in [0, D]$, we have $tA(a) + sA(b) \subset A(t^2 a + s^2 b)$. Hence, by the log-concavity of λ, the function under the integral sign, $f(a) = \lambda(A(a))$, possesses the property

$$f(t^2 a + s^2 b) \geq f(a)^t f(b)^s, \quad 0 \leq a, b \leq D, \ t, s > 0, t + s = 1. \tag{13.0.3}$$

The inequality remains to hold if we extend f by zero to the interval $(D, +\infty)$. Note also that $f(0) = 1$ and that f will be non-increasing and left continuous on $[0, +\infty)$. The last mentioned properties ensure that there exists a maximal number $m \in [0.D]$ such that $f(m) \geq \frac{1}{e}$. Hence,

$$\mathbf{E}\,\xi^{1/2} \geq \int_0^m f(a)\,da^{1/2} \geq \frac{1}{e}\,m^{1/2},$$

and

$$m \leq e^2 \left(\mathbf{E}\,\xi^{1/2}\right)^2. \tag{13.0.4}$$

On the other hand, let us apply (13.0.3) with $b = 0$, $a = \frac{a_0}{t^2}$, $r = \frac{1}{t^2}$ where $a_0 > m$ is a fixed parameter. Then we get

$$f(a_0 r) \leq f(a_0)^{\sqrt{r}} \leq e^{-\sqrt{r}}, \quad r > 1,$$

so

$$\mathbf{E}\,\xi = \int_0^{a_0} f(a)\,da + a_0 \int_1^{+\infty} f(a_0 r)\,dr \leq a_0 + a_0 \int_1^{+\infty} e^{-\sqrt{r}}\,dr = \left(1 + \frac{4}{e}\right) a_0.$$

Letting $a_0 \downarrow m$, we get $\mathbf{E}\,\xi \leq (1 + \frac{4}{e})\,m$ which together with (13.0.4) yields

$$\mathbf{E}\,\xi \leq (e^2 + 4e)\,(\mathbf{E}\,\xi^{1/2})^2.$$

This is just the desired estimate (13.0.1) with $C = \sqrt{e^2 + 4e}$.

Remark 13.2. The property (13.0.2) is very close to the description of χ_K^2 as a concave function. For example, in case of the unit Euclidean ball $K = B_2$, the function $\chi_K^2(x) = 1 - |x|^2$ is indeed concave. Khinchine-type inequalities for such functions were investigated by L. Berwald who showed (cf. [Ber] and [Bor1] for another proof), that for all $p > q > 0$, within the class of all convex bodies K in $\mathbf{R}^n$ and all concave functions g on K, the ratio

$$\frac{\|\,g\,\|_{L^p(K,\mu)}}{\|\,g\,\|_{L^q(K,\mu)}}$$

is maximized in case of the simplex $K = \{x \in \mathbf{R}_+^n : x_1 + \cdots + x_n \leq 1\}$ and the linear function $g(x) = x_1$. Here, μ denotes the uniform measure on K. In particular, this result implies a Khinchine-type inequality

$$\|\,g\,\|_{L^2(K,\mu)} \leq \sqrt{2}\,\|\,g\,\|_{L^1(K,\mu)}. \tag{13.0.5}$$

CHAPTER 14

Uniformly convex bodies

Let K be a symmetric about the origin, convex body in $\mathbf{R}^n$. We are going to examine the bound (9.0.9) for the Sobolev constant $s_n(K)$ with respect to the canonical metric generated by the norm

$$\|x\| = \min\{\lambda \geq 0 : x \in \lambda K\}, \quad x \in \mathbf{R}^n.$$

Being restricted to K, this metric is called the inner metric for K.

Note that in general, for any $x \in K$, the assumptions $x \pm u \in K$ imply $u \in K$, so $\chi_K(x) \leq 1$. Hence, by Theorem 9.2,

Proposition 14.1. *The Sobolev constant $s_n(K)$ defined with respect to the inner metric on K satisfies*

$$s_n(K) \geq c. \tag{14.0.1}$$

with some universal constant $c \geq \frac{1}{2}$ independent of K.

Actually, the bound (9.0.9) yields $c = \frac{1}{8}$, and the improvement can be reached using a different argument. Recall the classical Brunn-Minkowski inequality (cf. e.g. [B-Z]): for all $t \in (0,1)$ and all non-empty Borel sets $A, B \subset \mathbf{R}^n$,

$$\mathrm{vol}_n(tA + (1-t)B) \geq \left(t\, \mathrm{vol}_n(A)^{1/n} + (1-t)\, \mathrm{vol}_n(B)^{1/n}\right)^n.$$

Hence, the uniform probability measure μ on K possesses the same property:

$$\mu(tA + (1-t)B) \geq \left(t\, \mu(A)^{1/n} + (1-t)\, \mu(B)^{1/n}\right)^n.$$

Applying it to the couple (A, K) and to the couple $(K \setminus A, K)$, we get

$$\mu(tA + (1-t)K) \geq \left(tp^{1/n} + 1 - t\right)^n,$$

$$\mu(t(K \setminus A) + (1-t)K) \geq \left(tq^{1/n} + 1 - t\right)^n,$$

where $p = \mu(A)$, $q = 1 - p$. Adding the last two inequalities and letting $t \to 1$, in the limit we obtain:

Proposition 14.2. *For every Borel subset A of K of measure $\mu(A) = p$,*

$$\mu^+(A) \geq \frac{n}{2}\left(p^{\frac{n-1}{n}} + q^{\frac{n-1}{n}} - 1\right), \quad q = 1 - p. \tag{14.0.2}$$

Here, the perimeter is understood as in (9.0.1) with B_2 replaced by K. To be more precise, in the above argument one has to assume that A is smooth enough, and then one may easily extend (14.0.2) from smooth sets to all Borel measurable. In a slightly different form, the argument appeared in [B3]; the present formulation with similar proof is given by F. Barthe [Bar].

Letting $n \to \infty$ in (14.0.2) , we obtain a dimension free bound

$$\mu^+(A) \geq \frac{1}{2}\left(p \log \frac{1}{p} + q \log \frac{1}{q}\right), \quad q = 1 - p.$$

Actually, this inequality remains to hold for any log-concave probability measure concentrated on K ([B3]). Clearly, it implies that the Cheeger constant $c(K)$ is separated from zero; nevertheless, (14.0.2) contains more information concerning the behaviour of the isoperimetric function $I_\mu(p)$ near zero. For example, we have:

Corollary 14.3. *For any $n \in \mathbf{N}$*

$$\frac{n}{2} \leq \liminf_{p \to 0} \frac{I_\mu(p)}{p^{(n-1)/n}} \leq \limsup_{p \to 0} \frac{I_\mu(p)}{p^{(n-1)/n}} \leq n.$$

Here, the first inequality immediately follows from (14.0.2). In order to see the third one, it suffices to consider the set $A = p^{1/n}K$ in which case $\mu^+(A) = np^{(n-1)/n}$. Hence, $I_\mu(p) \leq np^{(n-1)/n}$, for all $p \in (0,1)$.

Proof of Proposition 14.1. We need to check that (14.0.2) implies the inequality (14.0.1) with constant $c = \frac{1}{2}$. Let $\alpha = \frac{n-1}{n}$ so that $n = \frac{1}{1-\alpha}$. Rewrite the (desired) inequality $\frac{1}{2(1-\alpha)}\,(p^\alpha + q^\alpha - 1) \geq cp^\alpha$ as

$$\frac{1}{2(1-\alpha)}\left(1 - \frac{1-q^\alpha}{p^\alpha}\right) \geq c, \quad 0 < p \leq \frac{1}{2}. \tag{14.0.3}$$

By direct differentiation, one sees that the function $p \to \frac{1-q^\alpha}{p^\alpha}$ is increasing on $(0,1)$. Hence, the left hand side of (14.0.3) is minimized as $p = \frac{1}{2}$, and we are reduced to $a_n \equiv n\,(1 - 2^{-1/n}) \geq c$. The sequence a_n is increasing, so $c = a_1 = \frac{1}{2}$.
Proposition 14.1 is proved.

As turns out, in spaces of large dimension, the Sobolev constant $s_n(K)$ can be much larger, and the inequality (14.0.1) can be considerably sharpened for "sufficiently" convex bodies K.

A convex, symmetric about the origin body K in $\mathbf{R}^n$ is called uniformly convex, if the corresponding modulus of convexity,

$$\delta_K(\varepsilon) = \inf\left\{1 - \left\|\frac{x+y}{2}\right\| : \|x\| = \|y\| = 1, \|x-y\| \leq \varepsilon\right\},$$

is positive for each $\varepsilon > 0$. Here as before, the norm in $\mathbf{R}^n$ is inner for K. The Banach space $(\mathbf{R}^n, \|\cdot\|)$ is said to have a modulus of convexity of power $p \geq 2$ with constant $C > 0$, if $\delta_K(\varepsilon) \geq C\varepsilon^p$, for every $\varepsilon \in (0,2]$. Equivalently (cf. e.g. [P]), for all $x, y \in \mathbf{R}^n$,

$$\frac{\|x\|^p + \|y\|^p}{2} - \left\|\frac{x+y}{2}\right\|^p \geq \kappa^p\,\|x-y\|^p, \tag{14.0.4}$$

for some $\kappa > 0$. For example (cf. [L-T]), the unit ℓ^p-ball $K = B_p$ with $p \geq 2$ satisfies (14.0.4) with optimal $\kappa = \frac{1}{2}$.

Under the assumption (14.0.4), the characteristic $\chi_n(K)$ may easily be estimated in terms of the constant κ. Indeed, for $x \in K$, let $x \pm u \in K$ with maximal possible value of $\|u\|$ (equal to $\frac{1}{2}\,\chi_K(x)$). Since $\|x \pm u\| \leq 1$, application of (14.0.4) to $x+u$ and $x-u$ (in the place of x and y, respectively) gives $1-\|x\|^p \geq \kappa^p\, 2^p \|u\|^p$. Consequently, we get:

Lemma 14.4. *Under* (14.0.4), *for all* $x \in K$,

$$\chi_K(x) \leq \frac{1}{\kappa}\,(1-\|x\|^p)^{1/p}. \tag{14.0.5}$$

To apply (9.0.9), we may use the above bound which leads to computing the norm $\|\, g\,\|_{L^{\frac{n}{n-1}}(K,\mu)}$ for the function $g(x) = (1-\|x\|^p)^{1/p}$. To a little simplify the task, let us note that g is concave on K and apply Berwald's inequality (13.0.5). With (14.0.5), we then get

$$\chi_n(K) \leq \frac{1}{\kappa}\,\|\, g\,\|_{L^{\frac{n}{n-1}}(K,\mu)} \leq \frac{\sqrt{2}}{\kappa}\,\|\, g\,\|_{L^1(K,\mu)}. \tag{14.0.6}$$

Now,

$$\int_K (1-\|x\|^p)^{1/p}\, d\mu(x) = \int_0^1 (1-t^p)^{1/p}\, dt^n = \frac{\Gamma(\frac{n}{p}+1)\,\Gamma(\frac{1}{p}+1)}{\Gamma(\frac{n+1}{p}+1)}.$$

Therefore, by (14.0.6) and Theorem 9.2,

$$s_n(K) \geq \frac{\kappa}{8\sqrt{2}}\, \frac{\Gamma(\frac{n+1}{p}+1)}{\Gamma(\frac{n}{p}+1)\,\Gamma(\frac{1}{p}+1)}. \tag{14.0.7}$$

One may also apply Stirling's formula, to develop the right hand side of (14.0.7):

Proposition 14.6. *Let K be a symmetric convex body in $\mathbf{R}^n$ satisfying the property* (14.0.4) *with parameters $p \geq 2$ and $\kappa > 0$. Then, with respect to the inner metric in K, the Sobolev constant satisfies*

$$s_n(K) \geq c_{n,p}\, n^{1/p}\, \kappa, \tag{14.0.8}$$

for some double indexed sequence such that $c_{n,p} \to c$, as $n \to \infty$ and $\frac{p}{n} \to 0$ (where c is a positive universal constant).

In particular, for unit ℓ^p-balls $K = B_p$ with $p \geq 2$, we have $s_n(B_p) \geq c_{n,p}\, n^{1/p}$.

CHAPTER 15

From Isoperimetry to LS$_q$-inequalities

To illustrate what can happen for l^∞ ball (when Dirichlet boundary are involved), we utilise an idea of [**Z3**] as follows. First we recall that for the nth Cartesian product of the unit intervals one has the following Classical Sobolev inequality with Dirichlet boundary conditions for the product measure ν_n on $[0,1]^n$.

$$\left(\int |f|^{\frac{n}{n-1}} d\nu_n\right)^{\frac{n-1}{n}} \leq \frac{1}{n}\sum_{i=1}^{n} \int |\nabla_i f|\, d\nu_n \tag{15.0.1}$$

Using (15.0.1), for any $q \in (1,\infty)$ with the help of Holder inequality, we obtain

$$\left(\int |f|^{\frac{n}{n-1}} d\nu_n\right)^{\frac{n-1}{n}} \leq \frac{1}{n^{1/q}} \int |\nabla f|_q\, d\nu_n \tag{15.0.2}$$

where $|\nabla f|_q$ denotes the l_q norm of the gradient. We note that if f is nonnegative, with $p \in (1,\infty)$ such that $\frac{1}{p}+\frac{1}{q}=1$, using the differentialtion rules and the Young inequality we get

$$\frac{1}{n^{1/q}}|\nabla f|_q = p^{\frac{1}{p}} f^{\frac{1}{p}} \cdot \frac{1}{n^{1/q}} q p^{-\frac{1}{p}} |\nabla f^{\frac{1}{q}}|_q \leq f + \frac{1}{n}(q-1)^{q-1}|\nabla f|_q^q$$

Inserting this into (15.0.2) and performing simple algebraic transformations, we obtain the following relation

$$\frac{\|f\|_{\frac{n}{n-1}} - \|f\|_1}{\frac{1}{n}} \leq (q-1)^{q-1} \int |\nabla f^{\frac{1}{q}}|_q^q\, d\nu_n$$

Hence passing to the limit with $n \longrightarrow \infty$ and replacing f by f^q, we arrive at the following LS$_q$-inequality

$$\mathrm{Ent}_\nu(f^q) \leq (q-1)^{q-1} \int |\nabla f|_q^q\, d\nu$$

with the infinte dimensional product measure ν on $[0,1]^{\mathbf{N}}$. (Note that the limiting procedure involves in fact also a sequence of functions, as we need to preserve the Dirichlet condition all time on the way.)

CHAPTER 16

Isoperimetric Functional Inequalities

The following functional inequality was first introduced in [**B2**] for the two point measure and via central limit theorem extended to the standard Gaussian measure

$$(\mathrm{IFI}_2) \qquad \mathcal{I}(\mu f) \leq \mu\left(\mathcal{I}(f)^2 + C|\nabla f|_2^2\right)^{\frac{1}{2}}$$

with the Gaussian isoperimetric function $\mathcal{I} \equiv \gamma(\Phi^{-1}(x))$, given by $\Phi'(x) = \gamma(x) \equiv \exp(-x^2/2)/\sqrt{2\pi}$ and some constant $C \in (0, \infty)$ for any function $0 \leq f \leq 1$ for which the right hand side is finite. One of the key properties of (IFI_2) is the product property which makes it suitable for the infinite dimensional analysis.

The result of [**B2**] was later extended in a number of directions including [**Ba-L**], [**B-G2**], [**Fo**], [**Z1**],... and others, expanding greatly the set of measures for which it holds true.

To recover the isoperimetric information from (IFI_2), we note that the function $\mathcal{I}$ is (concave, symmetric around $1/2$ and) vanishing at 0 and 1. Therefore one can use a suitable approximation of the characteristic function of a set A to obtain the following relation

$$(\mathrm{ISO}_2) \qquad \mathcal{I}(\mu(A)) \leq C\mu_2^+(A)$$

where the surface measure μ_2^+ corresponds formally to a suitable metric for which the natural length of the gradient of a Lipschitz function is $|\nabla f|_2$.

The inequality (IFI_2) is in general stronger than (ISO_2). However, following ([B2]) it is interesting to see by the following direct computation that using (ISO_2) with some extra information one can derive (IFI_2) in the Gaussian case. Namely if (ISO_2) is satisfied for the $\gamma^{\otimes 2}$, with $\gamma \equiv \gamma dx$ and $C = 1$, then for $A = \{x \leq \Phi^{-1}(f(y))\}$, defined with a smooth function $0 \leq f \leq 1$, one gets

$$\mu(A) = \int\int 1_{\{x \leq \Phi^{-1}(f(y))\}}\gamma(dx)\gamma(dy) = \int f(y)\gamma(dy) \tag{16.0.1}$$

and choosing natural parametrisation $(x, \Phi^{-1}(f(x)))$ of the boundary ∂A and using the definition of $\mu_2^+(\partial A)$ together with elementary calculus we obtain

$$\begin{aligned}
\mu_2^+(\partial A) &= \int \gamma\left(\Phi^{-1}(f)\right)\sqrt{1 + |\nabla\Phi^{-1}(f)|^2}\,\gamma(dx) \\
&= \int \mathcal{I}(f)\sqrt{1 + \frac{1}{\mathcal{I}(f)^2}|\nabla f|^2(x)}\,\gamma(dx) = \int \sqrt{\mathcal{I}(f)^2 + |\nabla f|^2}\,\gamma(dx) \qquad (16.0.2)
\end{aligned}$$

Thus the relations (16.0.1) and (16.0.2) together with (ISO) imply (IFI_2).

One of the key features of (IFI_2) is the product property which makes it suitable for infinite dimensional analysis. As shown in [**Z1**] the following natural generalisation of this inequality has also the product property

$$(\mathrm{IFI}_q) \qquad \mathcal{U}(\mu f) \leq \mu \left(\mathcal{U}(f)^q + C|\nabla f|_q^q\right)^{\frac{1}{q}}$$

with a (concave) function $\mathcal{U}$ defined on the unit interval and vanishing at 0 and 1, where $C \in (0, \infty)$ is independent of a function $0 \leq f \leq 1$ for which the right hand side make sense.

In this section we consider a function $\mathcal{U}(x) \equiv \varphi(F^{-1}(x))$ where F^{-1} is the inverse of the distribution function associated to the density $\varphi(z) = \alpha_r^{-1} e^{-|z|^r} dz \equiv \varphi_r(z)$ with $\alpha_r \equiv \int_{\mathbb{R}} e^{-|z|^r} dz$ being a normalisation factor and $r \in (1, \infty)$. The choice of $\mathcal{U}$ is motivated by the fact that in dimension one this is the right isoperimetric function for the measure $\nu_r(dz) \equiv \varphi_r(z)dz$, [**B-H2**]. Here, for $q \in (1, 2]$ we choose $r \in [2, \infty)$ such that $\frac{1}{r} + \frac{1}{q} = 1$ and show that in this case IFI_q implies LS_q, extending the known result for $q = 2$ (see e.g. [**Ba-L**];. we remark also that for that index $q = 2$ the converse implication was shown recently by the semigroup technique in [**Fo**]).

We begin from the following lemma in which (as well as later on) we use a simplified notation $\nu(dz) \equiv \varphi(z)dz \equiv \nu_r(dz)$.

Lemma 16.1.
(i) *For* $y < 0$, *we have*

$$r^{-1}|y|^{-r+1}(1 - q^{-1}|y|^{-r})\varphi(y) \leq F(y) \leq r^{-1}|y|^{-r+1}\varphi(y) \tag{16.0.3}$$

where $q^{-1} + r^{-1} = 1$.
(ii) *For* $x \neq \frac{1}{2}$, *we have*

$$\mathcal{U}'(x) = -r|F^{-1}(x)|^{r-1} sign(F^{-1}(x))$$

and

$$\mathcal{U}(x)\mathcal{U}''(x) = -\frac{r^2}{q}|F^{-1}(x)|^{r-2} \tag{16.0.4}$$

(iii)

$$\lim_{x \longrightarrow 0^+} \frac{\mathcal{U}(x)}{x \left(\log \frac{1}{x}\right)^{1/q}} = r \tag{16.0.5}$$

Proof of (iii). We note that

$$\lim_{x \longrightarrow 0^+} \frac{\mathcal{U}(x)}{x \left(\log \frac{1}{x}\right)^{1/q}} = \lim_{y \longrightarrow -\infty} \frac{\varphi(y)}{F(y) \left(\log \frac{1}{F(y)}\right)^{1/q}} \tag{16.0.6}$$

Using (16.0.3) and the fact that the function $x \left(\log \frac{1}{x}\right)^{1/q}$ is increasing for small x , we have

$$F(y) \left(\log \frac{1}{F(y)}\right)^{1/q} \leq r^{-1}|y|^{-r+1}\varphi(y) \left(\log \frac{1}{r^{-1}|y|^{-r+1}\varphi(y)}\right)^{1/q}$$

$$\leq r^{-1}|y|^{-r+1}\varphi(y)\left(|y|^r + \left|\log\frac{1}{r^{-1}|y|^{-r+1}\alpha_r^{-1}}\right|\right)^{1/q}$$

$$\leq r^{-1}\varphi(y)\left(|y|^{-r+1+r/q} + |y|^{-r+1}\left(\left|\log\frac{1}{r^{-1}|y|^{-r+1}\alpha_r^{-1}}\right|\right)^{1/q}\right)$$

Hence using the fact that $\frac{1}{r}+\frac{1}{q}=1$, we get

$$F(y)\left(\log\frac{1}{F(y)}\right)^{1/q} \leq r^{-1}\varphi(y)\left(1+|y|^{-r+1}\left(\left|\log\frac{1}{r^{-1}|y|^{-r+1}\alpha_r^{-1}}\right|\right)^{1/q}\right)$$

Thus

$$\frac{\varphi(y)}{F(y)\left(\log\frac{1}{F(y)}\right)^{1/q}} \geq r\left(1+|y|^{-r+1}\left(\left|\log\frac{1}{r^{-1}|y|^{-r+1}\alpha_r^{-1}}\right|\right)^{1/q}\right)^{-1} \longrightarrow_{y\longrightarrow\infty} r$$

Similar arguments using the lower bound (16.0.3) for the distribution F allows us to get the upper bound converging to r as y goes to $-\infty$. This ends the proof of (iii) .

□

Recall that given a probability measure μ and a nonnegative measurable function f we define the relative entropy as follows

$$\mathrm{Ent}_\mu(f) \equiv \mu\left(f\log\frac{f}{\mu f}\right)$$

Proposition 16.2. *Suppose* $r\in[2,\infty)$ *and* $\frac{1}{r}+\frac{1}{q}=1$. *Then there is a constant* $A\in[r,\infty)$ *such that the function*

$$\xi(x)\equiv \mathcal{U}^q(x)/x^{q-1} - Ax\log\frac{1}{x}$$

is convex on $[0,\frac{1}{2}]$ *and therefore for any probability measure* μ *and any measurable function* $0\leq f\leq\frac{1}{2}$ *we have*

$$\mathcal{U}^q(\mu f)/(\mu f)^{q-1} - \mu\left(\mathcal{U}^q(f)/f^{q-1}\right) \leq A\,\mathrm{Ent}_\mu(f) \tag{16.0.7}$$

Moreover, for any function $0\leq f\leq 1$, *we have*

$$\lim_{\varepsilon\longrightarrow 0^+}\frac{1}{\varepsilon}\left(\mathcal{U}^q(\varepsilon\mu f)/(\varepsilon\mu f)^{q-1} - \mu\left(\mathcal{U}^q(\varepsilon f)/(\varepsilon f)^{q-1}\right)\right) = r\mathrm{Ent}_\mu(f) \tag{16.0.8}$$

Proof. The inequality (16.0.7) follows by Jensen inequality from the convexity of the function ξ . To prove the convexity of ξ we note that

$$\xi'(x)\equiv q\mathcal{U}^{q-1}(x)\mathcal{U}'(x)/x^{q-1} - (q-1)\mathcal{U}^q(x)/x^q - A\log\frac{1}{x} + A$$

and hence

$$\xi''(x)\equiv q(q-1)\mathcal{U}^{q-2}(x)(\mathcal{U}'(x))^2/x^{q-1} - 2q(q-1)\mathcal{U}^{q-1}(x)\mathcal{U}'(x)/x^q$$
$$- q(q-1)\mathcal{U}^q(x)/x^{q+1} + q\mathcal{U}^{q-1}(x)\mathcal{U}''(x)/x^{q-1} + \frac{A}{x}$$

and therefore

$$\xi''(x) \equiv q(q-1)\mathcal{U}^{q-2}/x^{q-1}\left(\mathcal{U}'(x) - \mathcal{U}(x)/x\right)^2 + \left[q\mathcal{U}^{q-1}(x)\mathcal{U}''(x)/x^{q-1} + \frac{A}{x}\right] \tag{16.0.9}$$

Thus we need to show that the second term on the right hand side of (16.0.9) is nonnegative. To this end we note that, choosing $x = F(y)$ and using (16.0.4) we have

(16.0.10)

$$q\mathcal{U}^{q-1}(x)\mathcal{U}''(x)/x^{q-1} + \frac{A}{x} = q\mathcal{U}^{q-2}(x)(-r(r-1)|F^{-1}(x)|^{r-2})/x^{q-1} + \frac{A}{x} =$$

$$q\varphi(y)^{q-2}(-\frac{r^2}{q}|y|^{r-2})/(F(y)^{q-1}) + \frac{A}{F(y)} = \frac{1}{F(y)}\left[A - r^2|y|^{r-2}\varphi(y)^{q-2}/(F(y)^{q-2})\right]$$

Next, for $y \leq -1$ we use the lower bound from (16.0.3) to see that

$$\begin{aligned} r^2|y|^{r-2}/((r^{-1}|y|^{-r+1}(1-q^{-1}|y|^{-r}))^{q-2}) &= r^q|y|^{r-2+(r-1)(q-2)}/(1-q^{-1}|y|^{-r})^{q-2}) \\ &= r^q/(1-q^{-1}|y|^{-r})^{q-2}) \leq r^{2qr} \end{aligned}$$

where in the last line we have taken into the account that $(r-1)(q-1) = 1$ (because r and q are dual to each other). On the other hand if $r \geq 2$, one has that

$$\sup_{y\in[-1,0]} |y|^{r-2}\varphi(y)^{q-2}/(F(y)^{q-2}) < \infty$$

Hence choosing $A \equiv \max\left(r^{2qr}, \sup_{y\in[-1,0]} |y|^{r-2}\varphi(y)^{q-2}/(F(y)^{q-2})\right)$ we get $\xi(x) \geq 0$ for all $x \in [1, \frac{1}{2}]$.

The relation (16.0.8) follows from (16.0.7) and (16.0.5) .

Assuming we are given a natural gradient $\nabla f \equiv (\nabla_j f)_{j\in\mathcal{R}}$, with coordinates $\nabla_j f$ indexed by a countable set $\mathcal{R}$, we define its l_q norm $|\nabla f|_q$ as follows

$$|\nabla f|_q^q \equiv \sum_j |\nabla_j f|^q$$

Theorem 16.3. *Suppose $r \in [2, \infty)$ and $\frac{1}{r} + \frac{1}{q} = 1$ and let $\mathcal{U}(x) \equiv \varphi(F^{-1}(x))$ with $\varphi = \varphi_r = F'$. If there is a constant $C \in (0, \infty)$ such that*

$$\mathcal{U}(\mu f) \leq \mu\left(\mathcal{U}(f)^q + C|\nabla f|_q^q\right)^{\frac{1}{q}} \tag{IFI$_q$}$$

for any function $0 \leq f \leq \frac{1}{2}$ for which $|\nabla f|_q$ is integrable, then there is a constant $C' \in (0, \infty)$ such that

$$\mu f^q \log \frac{f^q}{\mu f^q} \leq C' \mu|\nabla f|_q^q \tag{LS$_q$}$$

for all nonnegative functions f for which the right hand side is finite.

Proof. Given a function f for which $|\nabla f|_q^q$ is μ integrable, we define a function $G(\lambda, z)$ of nonnengative arguments λ, $z \in [0, \infty)$, by

$$G(\lambda, z) \equiv \mu\left[\lambda^q f^q + z^q\ \left(\mathcal{U}(f)^q + C|\nabla f|_q^q\right)\right]^{\frac{1}{q}} - \left[\lambda^q\,(\mu f)^q + z^q\ \mathcal{U}(\mu f)^q\right]^{\frac{1}{q}}$$

with some $C \in (0, \infty)$. For later purposes we note that $G(\lambda, z)$ is continuous in its domain and uniformly continuous in an open neighbourhood of the origin. Its derivative with respect to z^q for $\lambda > 0$ is given by

$$\frac{\partial}{\partial z^q} G(\lambda, z) = \mu \frac{\left(\mathcal{U}(f)^q + C|\nabla f|_q^q\right)}{\left[\lambda^q f^q + z^q \left(\mathcal{U}(f)^q + C|\nabla f|_q^q\right)\right]^{1-\frac{1}{q}}} - \frac{\mathcal{U}(\mu f)^q}{\left[\lambda^q (\mu f)^q + z^q \mathcal{U}(\mu f)^q\right]^{1-\frac{1}{q}}}$$

In particular at $z = 0$ we have

$$\lambda^{\frac{q}{r}} \frac{\partial}{\partial z^q} G(\lambda, z = 0) = \left(\mu \left(\frac{\mathcal{U}(f)^q}{f^{q-1}} \right) - \frac{\mathcal{U}(\mu f)^q}{(\mu f)^{q-1}} \right) + q^q C \mu |\nabla f^{\frac{1}{q}}|_q^q \tag{16.0.11}$$

To show implication $\mathrm{IFI}_q \implies \mathrm{LS}_q$, we note that by Minkowski inequality one gets
(16.0.12)

$$G(\lambda, z) \geq \left[\lambda^q (\mu f)^q + z^q \left(\mu \left(\mathcal{U}(f)^q + C|\nabla f|_q^q \right)^{\frac{1}{q}} \right)^q \right]^{\frac{1}{q}} - \left[\lambda^q (\mu f)^q + z^q \, \mathcal{U}(\mu f)^q \right]^{\frac{1}{q}}$$

Thus if IFI_q is satisfied, the function $G(\lambda, z)$ is nonnegative and equal to zero for $z = 0$. This implies that its derivative with respect to z^q at $z = 0$ and $\lambda > 0$, is also nonnegative and using the formula (16.0.11) we have

$$\frac{\mathcal{U}(\mu f)^q}{(\mu f)^{q-1}} - \mu \left(\frac{\mathcal{U}(f)^q}{f^{q-1}} \right) \leq q^q C \mu |\nabla f^{\frac{1}{q}}|_q^q \tag{16.0.13}$$

We replace f by εf and divide both sides by ε . Passing with ε to 0 and using (16.0.8) we conclude that the following inequality is true

$$\mathrm{Ent}_\mu (f) \leq r^{-1} q^q C \mu |\nabla f^{\frac{1}{q}}|_q^q \tag{16.0.14}$$

For nonnegative functions f this is equivalent to LS_q . This ends the proof.

Bibliography

[A-M-S] Aida, S., Masuda, T., and Shigekawa, I., *Logarithmic Sobolev inequalities and exponential integrability.* J. Func. Anal. **126** (1994), 83-101.

[A-S] Aida, S., and Stroock, D., *Moment estimates derived from Poincaré and logarithmic Sobolev inequalities.* Math. Research Letters, **1** (1994), 75–86.

[A] Alesker, S., *Localisation technique on the sphere and the Gromov-Milman theorem on the concentration phenomenon on uniformly convex sphere.* Convex geometric analysis. (Berkeley, CA, 1996), 17–27, Math. Sci. Res. Inst. Publ. **34**, Cambridge Univ. Press, Cambridge, 1999.

[A-B-C-F-G-M-R-S] Ané, C., Blachére, S., Chafaï, D., Fougéres, P., Gentil, I., Malrieu, F., Roberto, C., and Scheffer, G., *Sur les inégalités de Sobolev logarithmiques.* (French) [Logarithmic Sobolev inequalities] With a preface by D. Bakry and M. Ledoux. Panoramas et Synthéses, **10**. Société Mathématique de France, Paris, 2000, xvi+217 pp.

[Ba-L] Bakry, D., and Ledoux, M., *Lèvy-Gromov Isoperimetric Inequality for an Infinite Dimensional Diffusion Generator.* Invent. Math. **123** (1996), 259-281.

[Bar] Barthe, F., *Log-concave and spherical models in isoperimetry.* Geom. Funct. Anal. **12** (2002), No.1, 32–55.

[Be] Beckner, W., *A generalized Poincaré inequality for Gaussian measures.* Proc. of the AMS **105** (1989), No.2, 397-400.

[Ber] Berwald, L., *Verallgemeinerung eines Mittelwertsatzes von J. Favard für positive konkave Funktionen.* Acta Math. **79** (1947), 17–37.

[B1] Bobkov, S. G., *Extremal properties of half-spaces for log-concave distributions.* Ann. Probab. **24** (1996), No.1, 35–48.

[B2] Bobkov, S. G., *An isoperimetric inequality on the discrete cube, and an elementary proof of the isoperimetric inequality in Gauss space.* Ann. Probab. **25** (1997), No.1, 206–214.

[B3] Bobkov, S. G., *Isoperimetric and analytic inequalities for log-concave probability measures.* Ann. Probab. **27** (1999), No.4, 1903–1921.

[B-G1] Bobkov, S.G., and Götze, F., *Exponential integrability and transportation cost related to logarithmic Sobolev inequalities.* J. Funct. Anal. **163** (1999), No.1, 1–28.

[B-G2] Bobkov, S.G., and Götze, F., *Discrete isoperimetric and Poincaré-type inequalities.* Probab. Theory Rel. Fields **114** (1999), 245–277.

[B-H1] Bobkov, S. G., and Houdré, C., *Isoperimetric constants for product probability measures.* Ann. Probab. **25** (1997), No.1, 184–205.

[B-H2] Bobkov, S. G., and Houdré, C., *Some connections between isoperimetric and Sobolev-type inequalities.* Mem. Amer. Math. Soc. **129** (1997), No.616, viii+111 pp.

[B-L] Bobkov, S. G., and Ledoux, M., *From Brunn-Minkowski to Brascamp-Lieb and to logarithmic Sobolev inequalities.* Geom. Funct. Anal. **10** (2000), No.5, 1028–1052.

[Bor1] Borell, C., *Complements of Lyapunov's inequality.* Math. Ann. **205** (1973), 323–331.

[Bor2] Borell, C., *Convex measures on locally convex spaces.* Ark. Mat. **12** (1974), 239–252.

[B-Z] Burago, Yu. D., and Zalgaller, V. A., *Geometric inequalities.* Springer Series in Soviet Mathematics, Springer-Verlag, Berlin, 1988. xiv+331 pp.

[C] Cheeger, J., *A lower bound for the smallest eigenvalue of the Laplacian.* Problems in analysis (Papers dedicated to Salomon Bochner, 1969), pp. 195–199. Princeton Univ. Press, Princeton, N.J., 1970.

[Cr] Croke, C. B., *Some isoperimetric inequalities and eigenvalue estimates.* Ann. scient. Èc. Norm. Sup., Ser. 4, **13** (1980), 419–435.

[D] Davies, E. B., *Heat kernels and spectral theory.* Cambridge University Press (1989).

[D-S] Davies, E. B., and Simon B., *Ultracontractivity and the heat kernel for Schrödinger operators and Dirichlet Laplacians.* J. Func. Anal. **59** (1984), 335–395.

[D-SC] Diaconis, P., and Saloff-Coste, L., *Logarithmic Sobolev inequalities for finite Markov chains.* Ann. Appl. Probab. **6** (1996), No.3, 695–750.

[F-G] Fradelizi, M., and Guédon, O., *The extreme points of subsets of s - concave probabilities and a geometric localization theorem.* Discrete Comput. Geom. **31** (2004), No.2, 327–335.

[Fo] Fougères, P., *Hypercontractivité et isopérimétrie gaussienne. Applications aux systèmes de spins.* Ann. Inst. Henri Poincaré, Probabilités et Statistiques 36, **5** (2000), 647-689.

[G-M] Gromov, M., and Milman, V. D., *Generalization of the spherical isoperimetric inequality to uniformly convex Banach spaces.* Compositio Math. **62** (1987), No.3, 263–282.

[G1] Gross, L., *Logarithmic Sobolev inequalities.* Amer. J. Math. **97** (1975), 1061–1083.

[G2] Gross, L., *Logarithmic Sobolev inequalities and contractivity properties of semigroups.* Varenna, 1992. Lecture Notes in Math. **1563** (1993), 54–88.

[G-R] Gentil, I., and Roberto, C., *Spectral gap for spin systems: some non-convex phase examples.* J. Func. Anal. **180** (2001), 66 - 84.

[G-Z] Guionnet A., and Zegarlinski B., *Lectures on Logarithmic Sobolev Inequalities*, IHP Course 98, pp 1-134 in Seminare de Probabilite XXVI, Lecture Notes in Mathematics **1801**, Springer 2003.

[H-S] Holley, R., and Stroock, D.W., *Logarithmic Sobolev inequalities and stochastic Ising models.* J. Stat. Phys. **46** (1987) 1159-1194 .

[K-L-S] Kannan, R., Lovász, L., and Simonovits, M., *Isoperimetric problems for convex bodies and a localization lemma.* Discrete Comput. Geom., **13** (1995), No.3-4, 541–559.

[K-A] Kantorovich, L.V., and Akilov, G.P., *Functional Analysis.* Pergamon Press, Oxford, 1982.

[L-O] Latala R. and Oleszkiewicz K., *Between Sobolev and Poincaré*, in Geometric Aspects of Functional Analysis, pp. 147-168, Lecture Notes in Math. **1745**, Springer, Berlin 2000.

[L1] Ledoux, M., *Isoperimetry and Gaussian Analysis.* Ecole d'été de Probabilités de Saint-Flour. (1994). Lecture Notes in Math. **1648** (1996), 165–294.

[L2] Ledoux, M., *Remarks on logarithmic Sobolev constants, exponential integrability and bounds on the diameter.* J. Math. of Kyoto Univ. **35** (1995), No.2, 211–220.

[L3] Ledoux, M., *On Talagrand's deviation inequalities for product measures.* ESAIM–Probability & Statistics, **1** (1996), 63–87.

[L4] Ledoux, M., *Concentration of measure and logarithmic Sobolev inequalities.* Séminaire de Probabilités, XXXIII, 120–216, Lecture Notes in Math. **1709**, Springer, Berlin, 1999.

[L5] Ledoux, M., *The concentration of measure phenomenon.* Mathematical Surveys and Monographs **89**. American Mathematical Society, Providence, RI, 2001, x+181 pp.

[L6] Ledoux, M., *Logarithmic Sobolev inequalities for unbounded spin systems revisited.* Séminaire de Probabilités, XXXV, 167–194, Lecture Notes in Math. **1755**, Springer, Berlin, 2001.

[Li] Li, P., *On the Sobolev constant and the p-srectrum of a compact Riemannian manifold.* Ann. scient. Éc. Norm. Sup., Ser. 4, **13** (1980), 451–469.

[Lie] Lieb, E.H., *Some convexity and subadditivity properties of entropy.* Bull. Amer. Math. Soc. **81** (1975), 1-13.

[L-T] Lindenstrauss, J., and Tzafriri, L., *Classical Banach spaces II*, Springer, 1979.

[L-S] Lovász, L., and Simonovits, M., *Random walks in a convex body and an improved volume algorithm.* Random Structures Algorithms **4** (1993), No.4, 359–412.

[M1] Maz'ja, V. G., *Classes of domains and embedding theorems for function spaces.* Dokl. Akad. Nauk SSSR **133**, 527–530 (Russian); translated as Soviet Math. Dokl. **1** (1960), 882–885.

[M2] Maz'ja, V. G., *Sobolev spaces.* Springer Series in Soviet Mathematics. Springer-Verlag, Berlin, 1985. xix+486 pp.

[Mu] Muckenhoupt, B., *Hardy's inequality with weights.* Studia Math. **XLIV** (1972), 31–38.

[N-S-V] Nazarov, F., Sodin, M., and Volberg, A., *The geometric KLS lemma, dimension free estimates for the distribution of values of polynomials and distribution of zeroes of random analytic functions.* Algebra i Analiz **14** (2002), No.2, 214–234 (Russian); English translation: St. Petersburg Math. J. **14** (2003), No.2, 351–366.

[P] Pisier, G., *Martingales with values in uniformly convex spaces.* Israel J. Math. **20** (1975), 326–350.

[Pr] Prékopa, A., *Logarithmic concave measures with application to stochastic programming.* Acta Sci. Math. (Szeged) **32** (1971), 301–316.

[Ro] Rothaus, O.S., *Analytic inequalities, isoperimetric inequalities and logarithmic Sobolev inequalities.* J. Func. Anal. **64** (1985), 296–313.

[Ros] Rosen J., *Sobolev Inequalities for Weight Spaces and Supercontractivity.* Trans. AMS **222** (1976), 367 – 376.

[V-P] Vauthier, J., and Prat, J.-J., *Cours d'analyse mathématique de l'agrégation.* (French) [Course on mathematical analysis for the Agregation]. Second edition. Agrégation de Mathématiques, Masson, Paris, 1994, viii+213 pp.

[Z1] Zegarlinski, B., *Isoperimetry for Gibbs Measures*, Ann. of Probab. **29** (2001), 802-819.

[Z2] Zegarlinski, B., *Entropy Bounds for Gibbs Measures with Non-Gaussian Tails*, J.Funct. Anal. **187** (2001), 368 - 395.

[Z3] Zegarlinski, B., *All that is Sobolev inequality*, unpublished.

Editorial Information

To be published in the *Memoirs*, a paper must be correct, new, nontrivial, and significant. Further, it must be well written and of interest to a substantial number of mathematicians. Piecemeal results, such as an inconclusive step toward an unproved major theorem or a minor variation on a known result, are in general not acceptable for publication. Papers appearing in *Memoirs* are generally longer than those appearing in *Transactions*, which shares the same editorial committee.

As of March 31, 2005, the backlog for this journal was approximately 7 volumes. This estimate is the result of dividing the number of manuscripts for this journal in the Providence office that have not yet gone to the printer on the above date by the average number of monographs per volume over the previous twelve months, reduced by the number of volumes published in four months (the time necessary for preparing a volume for the printer). (There are 6 volumes per year, each containing at least 4 numbers.)

A Consent to Publish and Copyright Agreement is required before a paper will be published in the *Memoirs*. After a paper is accepted for publication, the Providence office will send a Consent to Publish and Copyright Agreement to all authors of the paper. By submitting a paper to the *Memoirs*, authors certify that the results have not been submitted to nor are they under consideration for publication by another journal, conference proceedings, or similar publication.

Information for Authors

Memoirs are printed from camera copy fully prepared by the author. This means that the finished book will look exactly like the copy submitted.

The paper must contain a *descriptive title* and an *abstract* that summarizes the article in language suitable for workers in the general field (algebra, analysis, etc.). The *descriptive title* should be short, but informative; useless or vague phrases such as "some remarks about" or "concerning" should be avoided. The *abstract* should be at least one complete sentence, and at most 300 words. Included with the footnotes to the paper should be the 2000 *Mathematics Subject Classification* representing the primary and secondary subjects of the article. The classifications are accessible from `www.ams.org/msc/`. The list of classifications is also available in print starting with the 1999 annual index of *Mathematical Reviews*. The Mathematics Subject Classification footnote may be followed by a list of *key words and phrases* describing the subject matter of the article and taken from it. Journal abbreviations used in bibliographies are listed in the latest *Mathematical Reviews* annual index. The series abbreviations are also accessible from `www.ams.org/publications/`. To help in preparing and verifying references, the AMS offers MR Lookup, a Reference Tool for Linking, at `www.ams.org/mrlookup/`. When the manuscript is submitted, authors should supply the editor with electronic addresses if available. These will be printed after the postal address at the end of the article.

Electronically prepared manuscripts. The AMS encourages electronically prepared manuscripts, with a strong preference for $\mathcal{A}\mathcal{M}\mathcal{S}$-LaTeX. To this end, the Society has prepared $\mathcal{A}\mathcal{M}\mathcal{S}$-LaTeX author packages for each AMS publication. Author packages include instructions for preparing electronic manuscripts, the *AMS Author Handbook*, samples, and a style file that generates the particular design specifications of that publication series. Though $\mathcal{A}\mathcal{M}\mathcal{S}$-LaTeX is the highly preferred format of TeX, author packages are also available in $\mathcal{A}\mathcal{M}\mathcal{S}$-TeX.

Authors may retrieve an author package from e-MATH starting from `www.ams.org/tex/` or via FTP to `ftp.ams.org` (login as `anonymous`, enter username as password, and type `cd pub/author-info`). The *AMS Author Handbook* and the *Instruction Manual* are available in PDF format following the author packages link from `www.ams.org/tex/`. The author package can be obtained free of charge by sending email to `pub@ams.org` (Internet) or from the Publication Division, American Mathematical Society, 201 Charles St., Providence, RI 02904, USA. When requesting an author package, please specify $\mathcal{A}$$\mathcal{M}$$\mathcal{S}$-LaTeX or $\mathcal{A}$$\mathcal{M}$$\mathcal{S}$-TeX, Macintosh or IBM (3.5) format, and the publication in which your paper will appear. Please be sure to include your complete mailing address.

Sending electronic files. After acceptance, the source file(s) should be sent to the Providence office (this includes any TeX source file, any graphics files, and the DVI or PostScript file).

Before sending the source file, be sure you have proofread your paper carefully. The files you send must be the EXACT files used to generate the proof copy that was accepted for publication. For all publications, authors are required to send a printed copy of their paper, which exactly matches the copy approved for publication, along with any graphics that will appear in the paper.

TeX files may be submitted by email, FTP, or on diskette. The DVI file(s) and PostScript files should be submitted only by FTP or on diskette unless they are encoded properly to submit through email. (DVI files are binary and PostScript files tend to be very large.)

Electronically prepared manuscripts can be sent via email to `pub-submit@ams.org` (Internet). The subject line of the message should include the publication code to identify it as a Memoir. TeX source files, DVI files, and PostScript files can be transferred over the Internet by FTP to the Internet node `e-math.ams.org` (130.44.1.100).

Electronic graphics. Comprehensive instructions on preparing graphics are available at `www.ams.org/jourhtml/graphics.html`. A few of the major requirements are given here.

Submit files for graphics as EPS (Encapsulated PostScript) files. This includes graphics originated via a graphics application as well as scanned photographs or other computer-generated images. If this is not possible, TIFF files are acceptable as long as they can be opened in Adobe Photoshop or Illustrator. No matter what method was used to produce the graphic, it is necessary to provide a paper copy to the AMS.

Authors using graphics packages for the creation of electronic art should also avoid the use of any lines thinner than 0.5 points in width. Many graphics packages allow the user to specify a "hairline" for a very thin line. Hairlines often look acceptable when proofed on a typical laser printer. However, when produced on a high-resolution laser imagesetter, hairlines become nearly invisible and will be lost entirely in the final printing process.

Screens should be set to values between 15% and 85%. Screens which fall outside of this range are too light or too dark to print correctly. Variations of screens within a graphic should be no less than 10%.

Inquiries. Any inquiries concerning a paper that has been accepted for publication should be sent directly to the Electronic Prepress Department, American Mathematical Society, 201 Charles St., Providence, RI 02904, USA.

Editors

Titles in This Series

832 **Lee Klingler and Lawrence S. Levy,** Representation type of commutative Noetherian rings III: Global wildness and tameness, 2005

831 **K. R. Goodearl and F. Wehrung,** The complete dimension theory of partially ordered systems with equivalence and orthogonality, 2005

830 **Jason Fulman, Peter M. Neumann, and Cheryl E. Praeger,** A generating function approach to the enumeration of matrices in classical groups over finite fields, 2005

829 **S. G. Bobkov and B. Zegarlinski,** Entropy bounds and isoperimetry, 2005

828 **Joel Berman and Paweł M. Idziak,** Generative complexity in algebra, 2005

827 **Trevor A. Welsh,** Fermionic expressions for minimal model Virasoro characters, 2005

826 **Guy Métivier and Kevin Zumbrun,** Large viscous boundary layers for noncharacteristic nonlinear hyperbolic problems, 2005

825 **Yaozhong Hu,** Integral transformations and anticipative calculus for fractional Brownian motions, 2005

824 **Luen-Chau Li and Serge Parmentier,** On dynamical Poisson groupoids I, 2005

823 **Claus Mokler,** An analogue of a reductive algebraic monoid whose unit group is a Kac-Moody group, 2005

822 **Stefano Pigola, Marco Rigoli, and Alberto G. Setti,** Maximum principles on Riemannian manifolds and applications, 2005

821 **Nicole Bopp and Hubert Rubenthaler,** Local zeta functions attached to the minimal spherical series for a class of symmetric spaces, 2005

820 **Vadim A. Kaimanovich and Mikhail Lyubich,** Conformal and harmonic measures on laminations associated with rational maps, 2005

819 **F. Andreatta and E. Z. Goren,** Hilbert modular forms: Mod p and p-adic aspects, 2005

818 **Tom De Medts,** An algebraic structure for Moufang quadrangles, 2005

817 **Javier Fernández de Bobadilla,** Moduli spaces of polynomials in two variables, 2005

816 **Francis Clarke,** Necessary conditions in dynamic optimization, 2005

815 **Martin Bendersky and Donald M. Davis,** V_1-periodic homotopy groups of $SO(n)$, 2004

814 **Johannes Huebschmann,** Kähler spaces, nilpotent orbits, and singular reduction, 2004

813 **Jeff Groah and Blake Temple,** Shock-wave solutions of the Einstein equations with perfect fluid sources: Existence and consistency by a locally inertial Glimm scheme, 2004

812 **Richard D. Canary and Darryl McCullough,** Homotopy equivalences of 3-manifolds and deformation theory of Kleinian groups, 2004

811 **Ottmar Loos and Erhard Neher,** Locally finite root systems, 2004

810 **W. N. Everitt and L. Markus,** Infinite dimensional complex symplectic spaces, 2004

809 **J. T. Cox, D. A. Dawson, and A. Greven,** Mutually catalytic super branching random walks: Large finite systems and renormalization analysis, 2004

808 **Hagen Meltzer,** Exceptional vector bundles, tilting sheaves and tilting complexes for weighted projective lines, 2004

807 **Carlos A. Cabrelli, Christopher Heil, and Ursula M. Molter,** Self-similarity and multiwavelets in higher dimensions, 2004

806 **Spiros A. Argyros and Andreas Tolias,** Methods in the theory of hereditarily indecomposable Banach spaces, 2004

805 **Philip L. Bowers and Kenneth Stephenson,** Uniformizing dessins and Belyĭ maps via circle packing, 2004

804 **A. Yu Ol'shanskii and M. V. Sapir,** The conjugacy problem and Higman embeddings, 2004

803 **Michael Field and Matthew Nicol,** Ergodic theory of equivariant diffeomorphisms: Markov partitions and stable ergodicity, 2004

For a complete list of titles in this series, visit the AMS Bookstore at **www.ams.org/bookstore/**.